Global Warming and Climate Change

what Australia knew and buried

… then framed a new reality for the public

Global Warming and Climate Change
what Australia knew and buried
... then framed a new reality for the public

MARIA TAYLOR

PRESS

ANU PRESS

Published by ANU Press
The Australian National University
Canberra ACT 0200, Australia
Email: anupress@anu.edu.au
This title is also available online at http://press.anu.edu.au

---

National Library of Australia Cataloguing-in-Publication entry

Author:	Taylor, Maria.
Title:	Global warming and climate change : what Australia knew and buried ... then framed a new reality for the public / Maria Taylor.
ISBN:	9781925021905 (paperback) 9781925021912 (ebook)
Subjects:	Global warming--Australia. Climatic changes--Australia.
Dewey Number:	363.73874

Cover design and layout by ANU Press

Contents

Acknowledgements vii

The legacy ix

Foreword: the hidden history of Australia's early response
    to climate change xi

1. History is what we make it 1

2. Loading the dice: humans as planetary force 5

3. Framing information to influence what we hear 11

4. What Australians knew 25 years ago 17

5. Australians persuaded to doubt what they knew 45

6. Influences on a changed story and the new normal 1990s:
    values and beliefs 77

7. Influences on a changed story and the new normal:
    media locks in the new narrative 103

8. Influences on a changed story and the new normal:
    scientists' beliefs and public scepticism 133

9. In search of certainty and applying uncertainty 147

10. Dicing with the climate: How many more throws? 167

A chronology of some major climate science/policy milestones 181

List of acronyms 189

Bibliography 191

Acknowledgements

After many years spent as a journalist, editor and documentary film maker covering environmental and rural issues, I landed for an exciting spell of long-form research at the Australian National Centre for the Public Awareness of Science. My mission became to explore a hitherto unasked question that required ranging widely across science and the humanities: how could we, as a society, have buried what we once knew about the risk to everyone of global warming and climate change? This investigation was granted the necessary time through support from The Australian National University and the Commonwealth.

I am indebted to the journalists, scientists, policy-makers and researchers who put on the documentary record the outline of Australia's response to global warming and climate change since the 1980s. I am similarly indebted to those, particularly from the United States, who documented a parallel arc of public knowledge and official response over the past 25-30 years. The story that unfolded found strong echos in other countries with similar economics and politics.

You will find many names with attribution in the pages of this book. Their contributions to the national and international conversation are significant. Far from 'yesterday's news', they bore witness to today's history and how we communicated it along the way.

Public documents were the bedrock of this investigation and I am indebted also to the late Deni Greene and to the ACT Environment Centre for their super document collections showing Australia's early response to climate change knowledge. Thanks also to those who kindly granted interviews that expanded the documentary record. Finally, but not least, I thank my family, friends and professional colleagues who gave generously of their time, talent, ideas and encouragement to get this story told.

Maria Taylor

The legacy

Abysmal debate

The debate about the impact of human activity on climate change has been conducted on an abysmal level. The Rudd–Gillard–Rudd government comprehensively lost it by getting the politics wrong: failing to understand the fatal conjunction of inertia, self-interest, corporate power and media saturation. The relentless negativity and simplicity of the Coalition assertions, strongly supported by the Murdoch newspapers and shock-jocks on talk-back radio attacking the price of carbon ignored or derided the science and appealed to immediate economic self-interest.

Barry Jones (Commonwealth Science Minister 1983–1990), 'He did it his way to the end', *Canberra Times*, 15 November 2013, p. 4

Mother Nature will decide

Former US vice-president Al Gore:

'We have had deniers of the climate crisis in office in the US as well. History will not be kind to those who looked away, much less those who sought to prevent [action on climate change].'

Speaking of Australian Prime Minister Tony Abbott and his government's defunding of bodies to advise on and respond to climate change:

'I don't pretend to know what the basis of his thinking is, but Mother Nature has a louder voice' he said, referring to increasing incidences of severe weather.

'Al Gore: Mother Nature has "louder voice" than Tony Abbott', Nick O'Malley, *The Sydney Morning Herald*, 12 June 2014

Foreword: the hidden history of Australia's early response to climate change

Sydney will suffer twice as many days of extreme heat, four times as many severe storms and far worse flooding from huge increases in torrential rain, according to the latest predictions of how NSW will fare under the greenhouse effect.

These words were published in technology writer Gavin Gilchrist's 1995 article for the *The Sydney Morning Herald* headlined 'Greenhouse effect will cause havoc in NSW', which detailed a report by Australia's national, publicly funded, science organisation the CSIRO. At that time global warming/climate change was still called 'the greenhouse effect'.

The article warned of the increased risk of extreme heat and, therefore, fire, severe thunderstorms and torrential rains as the likely impacts of climate change. These are what we are coming to grips with globally as severe and catastrophic weather events. The article was far from the first on this topic.

As early as 1980 *Playboy* magazine published an indepth article on the subject quoting Australian scientists. *Newsweek* did a cover piece in 1987.

Hundreds of Australian mainstream and business press articles from the end of the 1980s into the early to mid-1990s provide a compelling record of how journalists and editors told the public about the risks posed by the greenhouse effect on weather events, on public health, on biodiversity, on Pacific islands, on Australian coastal communities and on society in general, including the likely impacts on gardening and holidaying in the Maldives.

These articles, as well as government documents and popular books published around this time, all readily ascribed the cause of the greenhouse effect to burning fossil fuels in industrial societies. The documentary record provides an indisputable and fascinating rendition of where we have been, how we thought and talked, what we once believed about global warming and climate change, and how that was reframed into a different story—all within 10 years, leaving us with the 'debate' we still struggle with today. (What was written also has the advantage of circumventing the limitations or preferences of human memory, which were evident in tracking down this story of our recent past).

I started reviewing the public record following disturbing reports of political interference in the communication of climate change science to the public, particularly in the United States. What was the situation in Australia, I wondered? Although I had been involved in the early 1990s in communicating some climate change-related energy efficiency strategies to the general public, by 2006 I was as clueless as the next person about what had happened to the climate change story. We had collectively lost the plot. I started to look at how and why.

We were ready to act in 1990 and called 'best informed' in the world

The evidence clearly shows that there has not been a one-way road of increasing scientific and public knowledge about global warming causes, climate effects and what societies can do. Indeed, the opposite has been the case in the two decades since the late 1980s.

Twenty-five years ago climate scientists spoke clearly and openly about global warming and the risks of climate change due to greenhouse gas emissions, particularly from burning fossil fuels. Leading politicians of both parties (yes, bipartisan), amplified by the media, repeated the messages of risk and vowed to act.

Following two major climate change conferences and community forums organised first by the CSIRO in 1987 and again, along with the federal Commission for the Future, in 1988, a study called the Australian public the best informed on the planet on this topic. This was also stated publicly at a United Nations' Global 500 Award ceremony during that period.

In October 1990, the federal Labor government under Prime Minister Bob Hawke established an interim emission reduction target for the nation to lower greenhouse gas emissions 20 per cent below 1988 levels by 2005.

Every state and territory drew up a detailed response plan. Every strategy that is known today to lower emissions, from efficiency and renewable energy to a carbon tax or price and emissions trading scheme was known then. The original Landcare one million tree planting program was started partly with the greenhouse effect in mind.

As late as the mid-1990s, technology and environment writers were still sounding the alarm, unequivocally. In 1996 Bob Beale wrote for the *The Sydney Morning Herald* about Australia's coal focus and its impact, asserting that it would take 420 million new trees to soak up the estimated 281 million tonnes of greenhouse gases related to the output of just one new Hunter Valley coalmine, according to government calculations. In context, in 2004 Australia's total greenhouse gas emissions as a country were 564 million tonnes, a figure that also shows the difficulties inherent in policies such as 'direct action' through tree planting.

The story for public consumption changed dramatically

While some reporters continued the story of risk to society, the overall narrative was changing dramatically by the mid-1990s. Communication from Australian policymakers, amplified by the media, had turned the story into a confused and conflicted political debate that reflected a loss of will to act. The state-based response plans would soon wither as deregulation focused the energy sector on competitive sales and profits rather than managing demand for efficient use.

Where once there had been a clear narrative about risk to the whole society and a global responsibility to act, Australians were now told not to worry: the whole 'debate' was too uncertain and prompt action was not in Australia's national and market interests.

All through this time, however, the scientific advice provided to governments by the UN Intergovernmental Panel on Climate Change (IPCC) since 1990 has remained consistent and unambiguous about causes, substantial risk and the need to respond sooner rather than later in order to slow down accumulating greenhouse gas emissions and their effect on the Earth's weather.

The science stayed on message, but our social reality and understanding were dramatically reframed as uncertainty in the 1990s. How that happened and what most influenced the story we as a society came to tell ourselves—the shift in cultural values, a different economic world view, the media's role in cementing a new narrative for the public to believe and, not least, the underrated importance of 'how' things were framed or said—including by scientists themselves—are brought together in this book.

If we don't understand where we have been, how public understanding can be reframed and manipulated and, indeed, how that was the story in Australia and in other Western democracies in the 1990s, it will remain easy to confuse the public and hard to move forward—as contemporary climate change politics continue to illustrate vividly.

The timeframe covered in this book is 1987 to 2001—starting before the first IPCC report in 1990 and then tracking print media stories and other public documents around the time of subsequent IPCC assessments that were released in 1995 and 2001. Interviews with scientists, reporters and policymakers on the scene at the time help to flesh out the details. The story is updated with examples from the 2000s and contemporary events, showing how the framing of communication to the public, manipulation of scientific findings and, most notably, the values and beliefs that defined much of the 1990s, continue to dominate (or attempt to dominate), national conversations on how to react to human-induced climate change.

One additional note is necessary. Recent research in the United States has shown that the terms 'global warming' and 'climate change' are often interpreted differently by different audiences. In this book I have largely used the term 'climate change' as shorthand for the enhanced greenhouse effect/global warming caused by human activities, leading to severe climate change.

1. History is what we make it

For almost 40 years I had the naive view that if we simply obtain more physical understanding of the issue, we could provide 'the' answers and responses would be rational. I now see that there is absolutely no guarantee of this. It is ourselves we do not understand.

Atmospheric scientist Graeme Pearman, 17 February 2009

As a top CSIRO climate scientist and head of his division from 1992–2002, Graeme Pearman contributed significantly to climate change knowledge internationally, as did his colleagues at the CSIRO Division of Atmospheric Research in Melbourne. He also was a tireless communicator to the public explaining the science and risks of climate change.

The fact that he found himself increasingly 'muzzled' in his public utterances by the late 1990s, and eventually made redundant from his position at CSIRO in 2004, reflects the trajectory of the climate change story in Australia during those years. Only now, more than a decade later, do climate scientists once more venture to publicly join the dots on why 'one-in-a-hundred-year' floods may be occurring every five to ten years and black Saturdays may well become more frequent.

Australia's recent history of climate change communication and understanding shows how we construct our own history—the stories we tell ourselves through the mass media and our beliefs about what is true or real can shift in as little as 10 to 20 years, and we then come to think things were always that way. Historians and philosophers know that social reality has shifted over time within our civilisation, and that it can be dramatically different from the world views of earlier civilisations. Psychology explains that knowledge is a social construct.

There are persuasive arguments that power comes from controlling the daily stories that define our society. And those who understand the cognitive sciences can tell us how related information frames and rhetoric are 'heard' and constructed. The narratives and agendas we listen to daily come from opinion leaders who are mostly politicians, corporate chieftains and the mass media (Lakoff 2004; Rampton & Stauber 2002; Ward 2001).

With that understanding, one can start unpacking the documentary record of how anthropogenic climate change was communicated from the late 1980s to the 2000s in Australia—and arrive at a startling discovery. We find that Australia once had a high level of understanding of global warming and climate change risks and possible responses, which was conveyed by scientists like Pearman, both here and internationally, and amplified by journalists and book authors, with not a sceptic in sight, at first. Polls indicate the general public wanted an active response. State and federal policymakers were persuaded of an urgent need to take action and started the wheels turning before 1991 (Henderson-Sellers 1990; Lowe 1989; Bulkley 2000b).

By 1990, Australia had an early national emissions reduction target—aiming for 20 per cent below 1988 emissions levels by 2005. Interviews indicate almost no-one remembers this, including some of those in government at the time, but the documentary record, particularly the newspaper record, is clear: 25 years ago Australia was agreeable to a tough emission reduction target.

In the late 1980s and early 1990s human responsibility for the enhanced 'greenhouse effect' was a given in every press article reviewed as well as in government documents. Response measures started with energy efficiency and canvassed every response—including a carbon price—that are discussed today. But as the 1990s progressed, this changed. Gradually the early good understanding was overtaken by a huge case of uncertainty, doubt or ignorance, which was closely linked to a changed daily narrative for public consumption that was crafted by political leaders and the mass media. A new normal was being created. What happened and how is the question I delve into in this book.

Equally remarkable is the fact that the science information about causes, effects and risks has remained consistent over nearly three decades. Every five to six years, mass media reports on a new international assessment that makes it sound like we have just confirmed that human activities are responsible, but this is hardly the case. One of the most emphatic assessments by the United Nations Intergovernmental Panel on Climate Change (IPCC) was the first it released in 1990. Assessments have followed in 1995, 2001, 2007 and 2013 (www.ipcc.ch/) and involve hundreds of the world's experts in their field analysing the growing body of science, economic impact and response research.

The 1990 report made the strongest and most easily understood (by the layperson) statements on the issue: climate change is happening and human burning of fossil fuels is the main agent. Since then, the most significant changes to IPCC assessments have been the inclusion of more localised detail, plus the realisation that the planet is experiencing an unsettling and unexpected rate of rapid climate change.

The evidence trail shows, however, that by 1996 and thereafter, with basically the same science story as laid out in the first IPCC report in 1990 (albeit with changed communication style), risk messages were being reframed into a hazy scientific debate, particularly about human agency, that confused the public and helped those who blocked action. The narrative that once asked what could be done to slow or reverse the emission of excess greenhouse gases by human societies, i.e. an early risk management and global ethical argument, evolved into an inward-focused national interest argument for no change from 'business as usual'.

The science story became an economic story and the storyline became a familiar contemporary one: Australia is exceptional amongst countries, thanks to policy decisions to focus the national economy on mineral and coal exports, and 'cheap' electricity production for the domestic market and to attract energy-intensive multinational industries like aluminium. With this narrative, Australia was reconstructing its social reality in the 1990s.

How the story was being revised first emerged from whistleblowing about political interference in science communication, which was extensively documented by the Union of Concerned Scientists (UCS) in the United States (http://www.ucsusa.org/global_warming/). In 2007 two further and complementary investigations in the United States confirmed these findings: there has been broad interference in the United States in the communication of scientific results relating to climate change. Interference came in the form of reports being altered or shelved, scientists harassed by Congressional committees and pressure to eliminate words like 'climate change' from research conclusions ('Dirty tricks' 2007; UCS 2007).

In Australia, while there was less documentation of direct political interference in communication in the past 20 years, a long-time chronicler of the climate change policy story, physicist and science and society researcher Ian Lowe wrote that it is done by appointment: 'the stacking and sacking of public boards, reviews and task forces has been driven by ideology and is suppressing new ideas arising from science, to the detriment of innovation and the environment' (Lowe 2006: 41).

Unveiling how communication has been manipulated over two decades may well hold the key to understanding why so many still don't understand. In terms of *why* the communication was reframed and manipulated, one cannot underestimate the influence of ideas that gained cultural dominance during the 1990s following a brief attempt to reconcile environmental and economic values up to 1991. The upsurge of economic market fundamentalism, in tandem with a return to a familiar battle pitting the economy *against* environmental ideas and environmental science, came to dominate policy responses.

Other dominant values and beliefs that surfaced to drive the argument against climate change action included traditional assumptions about human and Christian exceptionalism, and beliefs in salvation through technology and the 'techno fix'. This suite of values—inhabiting political, business, and some religious opinion leaders and amplified by the media—is shared by many in society at large.

It's a consistent world view, allowing those who hold these values to dismiss the risk or misread the science. Scientists as a class are not immune to these value structures. The public sceptics who helped foster uncertainty since the 1990s might well hold some such beliefs while also holding ideas particular to their scientific disciplines, geology being one example.

The gauntlet was thrown down as scientific sceptics (coming from any discipline), policymakers, media and the public grappled for the first time with a radically new concept: the notion of anthropogenic climate change—the idea that humans are now a force of nature capable of altering basic earth systems, such as the atmosphere.

2. Loading the dice: humans as planetary force

What we see happening with new record temperatures, both warm and cold is in good agreement with what we predicted in the 1980s when I testified to Congress about the expected effect of global warming. I used coloured dice then to emphasize that global warming would cause the climate dice to be 'loaded'—for risk of more extreme weather.

James Hansen, Director, Goddard Institute for Space Studies, interview with Bill McKibben, 22 December 2010

The current rate at which CO_2 is rising, 2 ppm per year, is unprecedented in the recent history of the Earth, with the exception of the onset of greenhouse atmospheric conditions following major volcanic episodes and asteroid and comet impacts, which led to the large mass extinctions [and ended planetary periods like the Jurassic and Cretaceous].

Andrew Glikson, *The lungs of the Earth*, countercurrents.org, 2009

The idea that the human species and its societies are a new 'force of nature' capable of altering planetary systems is a recent one that confronts long-held beliefs. That we are now in a new epoch called the Anthropocene is still resisted by some traditionally trained geologists and meteorologists, among others, and this has had implications for present-day sceptic debate.

How we got to this understanding takes us along the global warming/climate change science discovery path. Although there were earlier relevant discoveries, the path is generally described as starting with the 1890s hypothesis of Swedish scientist Svante Arrhenius that gases from burning fossil fuels could raise global temperatures. From then a complex, multidisciplinary research effort has led to the present-day understanding that human activities are changing the atmosphere–ocean–biosphere balance resulting in the warming greenhouse effect on the planet. Science historian Spencer Weart noted that at the turn of the 20th century, and for a long time thereafter, deeply embedded in human culture was the belief that either God or nature would take care of any human impacts:

> Hardly anyone imaged that human actions, so puny amongst the vast natural powers, could upset the balance that governed the planet as a whole ... It was traditionally tied up with a religious faith in the God-

given order of the universe ... Such was the public belief and scientists are members of the public, sharing most of the assumptions of their culture. (Weart 2004:8)

Not only was it considered unlikely that humans could affect earth systems, the *rate* of change indicated by climate models went counter to long-held beliefs and principles—promoted particularly by geologists who had a century earlier explained the phenomenon of coming and going ice ages for a disbelieving scientific community.

The discipline of geology holds that changes to planetary systems and climate can only occur as they have before (and, indeed, there have been many previous hotter and colder periods) which can be read from geological evidence researchers could measure on the ground. According to geologists, previous climatic changes occurred over thousands, if not millions, of years, which belief prompted a basic scepticism about rapid change induced by human activity.

During the second half of the 20th century other scientists were looking at planetary systems in detail—including the capacity of the oceans to absorb carbon dioxide (CO_2), thus delaying measurable on-land impacts for additional decades—and they were steadily learning about the connections between the world's biomass and ecosystems. Examples are the Arctic tundra as a reservoir of methane that would be released with warming, or the weakening of the Atlantic Gulf Stream leading to paradoxical cooling in the north Atlantic.

Earth scientists of all stripes only gradually learned that climate could change rapidly in just the span of a hundred years, or even a decade, and not solely over thousands of years or geological periods as previously thought. This understanding came with the disturbing corollary that rapid climate change might manifest no differently in the first instance than natural variation.

Rapid climate change and detecting the human fingerprint

How fast can our planet's climate change? Too slowly for humans to notice, was the firm belief of most scientists through much of the 20th century ... Today, there is evidence that severe change can take less than a decade. A committee of the (US) National Academy of Sciences (NAS) has called this reorientation in thinking of scientists a veritable 'paradigm shift' ... but this new thinking is little known and scarcely appreciated in the wider community of natural and social scientists and policymakers. (Weart 2004)

The same process of discovery was leading climate researchers to the conclusion that it was indeed human agency changing the atmosphere and the climate in this period of history. As early as the 1950s, US oceanographer Roger Revelle, who was studying the uptake of CO_2 in the oceans and CO_2 emission from industrial processes, came to a radical conclusion: 'Human beings are carrying out a large-scale geophysical experiment of a kind that could not have happened in the past nor be reproduced in the future' (Weart 2004: 30).

Neither Revelle nor other researchers then foresaw just how this experiment would ramp up as both industrialisation and population exploded during the next 50 years, accelerating the level of greenhouse gas emissions accumulating in the atmosphere as well as other significant and sometimes related impacts on earth systems.

Building on this background of 20th century research, ecologists and other scientists studying global change during the past decades under the International Geosphere Biosphere Project (IGBP) have been publishing the evidence for an 'Anthropocene epoch'; that is, the beginning of a time span where humans are the main planetary force altering natural systems. 'Anthropogenic climate change' refers to this human agency.

The IGBP dissected the cumulative human impact on the previous balance within the natural systems of soil, air, water, forests and species. Australia became involved in the project in 1990. In 2008 IGBP alumnus Will Steffen, former director of The Australian National University Climate Change Institute and recent member of the now disbanded federal Climate Commission, gave a seminar that summarised why the IGBP scientists believed that they had identified a new but massive human footprint over all earth systems. These scientists proposed a reconceptualisation of history, which tracks the evolution of modern societies against natural system benchmarks—including CO_2 in the atmosphere (Steffen, Crutzen & McNeill 2007; Costanza, Graumlich & Steffen 2007). The story they present in regard to the greenhouse effect goes as follows.

Stability for about 250,000 years

About 250,000 years ago, fully modern humans emerged in Africa. At that time, the concentration of CO_2 in the atmosphere was low—somewhere below 200 ppm—compared with today's 400 ppm. Atmospheric methane was similarly low. The concentration of both these gases rose for centuries at a time (but not above 240 ppm) and then fell for longer periods of time. This pattern steadied at 240 ppm from the beginning of agriculture, 5,000–7,000 years before the present, and through the great European civilisations of Greece and Rome.

Early human activities that may have contributed to relatively small elevated levels of CO_2 included fire-stick farming and forest clearing. Evidence for these conclusions come from Greenland ice cores.[1] The dramatic increases in CO_2 levels, however, started with the Industrial Revolution as can be seen in this graph.

The Stages of the Anthropocene

Pre-Anthropocene events:
Fire-stick farming, mega fauna extinctions, forest clearing

Anthropocene Stage 1 (ca 1800–1945)
Internal combustions engine, fossil fuel energy, science & technology

Anthropocene Stage 2 (1945–2010 or 2020)
The Great Acceleration, new institutions and vast global networks

Anthropocene Stage 3 (2010 or 2020–?)
Sustainability or collapse?

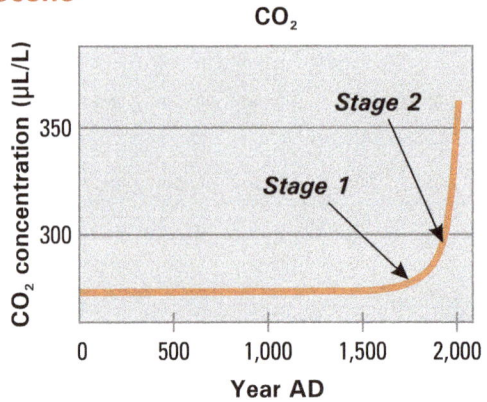

From: Steffen, Crutzen & McNeill 2006

Proxy evidence from 250,000 BC to the 1800s indicates natural systems remained remarkably stable in greenhouse gas concentrations until the beginning of the Industrial Revolution.

Source: Steffen, Crutzen & McNeill 2006.

CO_2 levels remained fairly constant at approximately 270 ppm or lower until the beginning of the Industrial Revolution. Starting around 1800, came the invention of the internal combustion engine, using fossil fuel energy, and what we call 'progress' made possible by modern science and technology. CO_2 levels started climbing slowly. During the period called 'Anthropocene stage 2' (1945 to 2010 or 2020), however, CO_2 levels rose rapidly and are still climbing (390 ppm at the end of 2009; 400 ppm in 2013).

The IGBP researchers found a matching curve of impacts in a range of biophysical reactions—away from the relative stability of previous centuries in human

---

1 Ice core data have been collected by the US National Oceanic and Atmospheric Administration (NOAA), among other expert worldwide agencies. Australian scientists pioneered extracting air from ice cores to 'read' prior history of atmospheric concentrations of greenhouse gases. NOAA's website states that data from polar and mountain glaciers and ice caps are archived yielding 'proxy' climate indicators from the past in oxygen isotopes, methane concentrations, dust content and other parameters (http://www.ncdc.noaa.gov/paleo/icecore.html).

history. Greenhouse gases, ozone depletion, full exploitation of fisheries, loss of forests, water shortages and species extinctions (among other benchmarks) zoomed upwards in just over 50 years.

The net result of maximised resource exploitation with industrialisation and population growth is described by the metaphor of a 'global footprint'. The implication is that humans are depleting the natural capital of the planet at an unsustainable rate in comparison to how many earths we would need to keep up with our demands (more than one additional earth by 2001).

Along with humans depleting the natural physical environment has come another startling and depressing realisation: humans are now seen by many scientists as the force that will lead to the sixth great extinction of other species—akin to the impact of a comet strike, or a series of volcanic eruptions that blot out the sun for years, or a natural overdose of CO_2 (which happened 252 million years ago at the end of the Permian) along with related impacts (also happening now), such as ocean acidification. The Permian climate change happened very rapidly in geologic terms.

> Temperatures soared—the seas warmed by as much as 10 degrees—and the chemistry of the oceans went haywire, as if in an out-of-control aquarium. The water became acidified, and the amount of dissolved oxygen dropped so low that many organisms probably, in effect, suffocated. Reefs collapsed. (Kolbert 2014: 103)

Interestingly, the study of previous extinctions has also suggested that the Anthropocene might have started much earlier—with the extinction of the megafauna and resultant vegetation changes on various continents as humans took up habitation (Kolbert 2014).

In similar fashion, by burning fossil fuels and other greenhouse gas-emitting activities, humans are now changing or altering climatic patterns—with impacts we can witness—such as increased cyclone force and frequency, increased flooding, and the heatwaves of the past decades, all set to accelerate. So-called 'one-in-a-hundred-year' intense bushfire and extensive flooding have followed in quick succession, and have decimated communities and regions. Steffen and his co-researchers postulate that, by 2050, heatwaves (and their flow-on effects) will be an everyday event.

Some theorists about Earth climate cycles see a paradox here. They argue that the rise in CO_2 and other greenhouse gases, prompting higher planetary temperatures (global warming), currently shield human civilisation from the natural climate pattern of the past millions of years. The dominant pattern is ice ages broken by short interglacial periods. Agricultural civilisations emerged in a balmy interglacial that started about 20,000 years ago, and some postulate that the Earth is due for another ice age. The problem for this theory is that it doesn't help on the ground: the bumpy ride predicted with climate change already in motion cancels any modifying influence to stave off ice ages in a way that matters to current civilisations.

Of more immediate consideration, humans appear hardwired for short-term thinking. Psychological concepts of how we view the world around us, including 'creeping normalcy' or 'landscape amnesia', block day-to-day comprehension of what accelerating human activities represent—whether it is human population, the number of dammed rivers, forest destruction, or the impact of motor car emissions in a timespan that is geologically brief. Creeping normalcy refers to slow trends concealed in noisy fluctuations that people get used to without comment, while landscape amnesia describes forgetting how different the landscape looked 20–50 years ago (Diamond 2005: 425).

In his study of how societies fail, biogeographer Jared Diamond calls global warming a pre-eminent example of a 'slow trend concealed by wide up and down fluctuations' (2005: 425). He likens the denial of climate change impacts by leading politicians, including former US president George W. Bush (and his contemporary John Howard in Australia), in the late 1990s and early 2000s to the elite of 'the medieval Greenlanders [who] had similar difficulties recognizing that their climate was gradually becoming colder, and the Maya and Anasazi (in Central and North America) [who] had trouble discerning that theirs was becoming drier' (2005: 425).

The evidence that humans have become a geophysical force is compelling. But denying or not acknowledging this was a hallmark of the framing of climate change in Australia from at least the mid-1990s. Short-termism and, more importantly, a shift in dominant values shaped that response, as we will see in upcoming pages. But a look first at the dynamics of 'what I say and what you hear' helps set the outline of how communication works in practice and suggests how it can be manipulated.

3. Framing information to influence what we hear

… perceptions are shaped not only by scientists but by interest groups, politicians and the media …

… the climate in the future actually may depend on what we think about it … what we think, will determine what we do …

Spencer Weart, 'The discovery of rapid climate change', 2004

If humans have become a geophysical force, then physicist Spencer Weart's history of climate change discovery points out just how important communication is to the way humans influence their biophysical surroundings based on what they believe to be true. What we think filters and translates a scientific message. Or rather, in Western democracies, it is what politicians, the media and the blogosphere think and say that influences what the general public thinks.

The evidence for Australia shows that the dominant narrative about the greenhouse effect/global warming/climate change was altered dramatically from how it started in 1987 through 1991, despite the consistency of the scientific message. How did interest groups, politicians and the media disengage public knowledge from the scientific facts?

The importance of language and framing are at the fore in free market driven democracies like Australia or the United States where regulation is disdained and, therefore, every citizen, or at least a voting majority, must be convinced of the need to act. Should we talk about global warming or climate change (or both tied together)? Does it matter? Apparently so, according to a report on recent nationally representative surveys conducted by the US Yale Project on Climate Change Communication and the George Mason University Center for Climate Change Communication (Leiserowitz 2014).

Researchers there have concluded that the terms 'global warming' and 'climate change', at least in the United States, garner significantly different responses from sections of the public, and that those differences relate to what different ethnic groups in society hear and respond to. 'Global warming' reportedly resonates for many with images of melting ice flows and weather catastrophe, while 'climate change' sounds more benign and is more easily accepted as a natural phenomenon.

In the United States, the change to the common use of 'climate change' to characterise the enhanced greenhouse effect was apparently also deliberately political and made on public relations advice that was received by the George W. Bush administration (Goldberg 2014). Scientists also prefer climate change as a more technically accurate term when a short, lay description is required.

What to do? In this book I have largely used the term climate change as shorthand for the enhanced greenhouse effect or global warming and severe climate change. It is not an either or proposition. Perhaps it is a matter of being aware of what audience is being addressed in the fragmented information environment of today. These findings underline the critical importance of how environmental and many public interest issues are framed and what people are likely to hear and act upon.

The fields of psychology, educational theory, linguistics and neuroscience illuminate the ways in which information is framed and the mental pathways of understanding. They tell us that it is not what you 'say' that matters, but what people 'hear'. This has underpinned education and learning theory for decades and, since the turn of the century, this understanding has been put to good use by propagandists and, more recently, by the public relations industry. This way of looking at how people take up information tells us that people hear or process information and interpret reality based on the sum of their past experience, not least through the filters of professional training as well as their core values, including religion (Lakoff 2004).

In other words, everyone reacts to information based on their memory banks and emotional triggers, which helps explain why communication can be so puzzlingly difficult. Furthermore, there is also a cultural creation of 'reality' beyond physical assumptions about solid matter. Many cultural edicts come from economic theories and religion. Within a given culture there is, then, unspoken agreement to perceive the world in certain ways, or 'myths to live by' in the words of historical philosopher Ronald Wright (2005). So processing science information is not a straightforward path.

Our neurological, social, and cultural constructions of reality dictate what we think about the world around us. This can easily lead to confusion and effective scare campaigns when manipulated, as the climate change story has been: for example on the matter of what action will 'cost' the public.

How we use language and metaphor is a basic part of constructing our reality. Most people probably have not thought of metaphors since high school English, but these mental pathways drive much of our language use. For example, in Western culture there is the influential metaphor of life as a journey: people must have a purpose—'find their way', have 'goals to reach' or they are 'lost', 'without direction'; those who 'reach their goals' fastest are admired, or maybe they have to 'find a different path'.

Cognitive scientist George Lakoff applied this understanding to techniques used in politics and by public relations practitioners to reinforce certain ideas and values. A common technique is to tap into a metaphorical pathway that has positive emotional values, such as freedom, home or family, in order to frame how to think about things (Lakoff 2004).

'Freedom' is a classic example of a metaphorical pathway that evokes everything we hold dear about our way of life—as shorthand, the concept merges political, economic and cultural aspirations. Everything attached to the word 'freedom' can evoke positive emotions and advertising has long exploited this understanding.

In the late 1960s, University of California communications professor Herbert Schiller wrote about the connection between mass media and American-style commerce and consumption, which is framed as the presence of freedom—in trade, speech and enterprise. In the war of ideas that accompanied the resurgence of neo-liberal economics since the 1970s, this also came to include freedom *from* government regulation of business, a perspective applied by politicians to environmental or public health issues—both relevant to climate change—that might otherwise have invited regulation (in Wheelright 1987).

National interest, jobs, family, battlers, Australian working families, Australian mums and dads: these terms and phrases are intended to evoke a framework of emotional responses. Family and jobs and country are cross-cultural themes and most Australians can understand or respond to these emotional levers. By linking 'jobs' and 'family', 'national interest' and 'responsible science', or 'needing more research' to messages about delaying action or challenging the science of climate change, members of the public may be induced to change their understanding—forgetting that once they were responding to frames about risk insurance, win-win energy policies, and responsible global citizenship.

It gets more complicated still. Environmental science messages are not only filtered through the use of language and individual understanding; information flow is also affected by institutional interactions (e.g., between researchers and politicians) and also by the influence of ideas and values that are dominant in society. Politics and the media are two primary institutions in the framing of science messages, along with scientists' own disciplinary cultures, as can be traced in the climate change story in Australia and other Western democracies.

An excellent example is the work of US political consultant and pollster Frank Luntz. During the past two decades he has developed manuals for conservative politicians that tell them what to say to have the desired impact, and he has advised particularly about climate change. In a key memo on climate change, called 'winning the global warming debate' Luntz (2003) advised that a primary strategy was to stoke the fires of scientific uncertainty, and to have scientists do the stoking. Voters had to be persuaded that there was no consensus.

The narrative had to be about us and them, differentiating the mainstream from environmental activists and by implication pitting our family, our nation, against 'them'. Other themes were: 'the right decision, not a quick decision'; 'voluntary innovation and experimentation are preferable to bureaucratic or international intervention and regulation'; 'fairness'—why should we take action and not those other countries?

Luntz feared that, should people come to believe that climate science is settled, they would want to act accordingly and demand action from their governments. Wanting action was indeed how Australia appeared to react around 1990 when the dominant narrative proposed that the science was clear-cut and the government opted for a vigorous response. Looking over Luntz's recommendations, one can see a playbook for the reframed story that occurred in Australia during the later 1990s and into the 2000s.

On 13 July 2009 the Australian Broadcasting Corporation (ABC) *Lateline* program featured the author of a recent book on climate change politics, (Lord) Anthony Giddens from the London School of Economics. He showed how successful this approach of promoting uncertainty was, telling the interviewer that in his surveys of populations in different countries an average 40 per cent of respondents were sceptical that scientists agreed about anthropogenic climate change. The actual case, said Giddens, was that perhaps one per cent of scientists working in the field of climate change remained sceptical of the general message.

The mass media and what they hear

In Australia, as in the United States, the public gets most of its science information through the media (Denemark 2005; Russell 2006), which means that what the media choses to hear and how it frames what it puts out is key to what the public will perceive as the facts on a science subject. In Australia's concentrated media market, Rupert Murdoch's News Limited has a history over the past 20 years of framing climate change science and responses as uncertain and unnecessary, while other print media has changed its narrative over time.

The delivery style also helps to frame the story. Marshall McLuhan, the Canadian critic of mass communications, said after World War II that the mass media shapes our daily narrative of reality by delivering drama, featuring us and them, good and bad, winners and losers (McLuhan 1951). This has not changed.

Research on how public agendas are set in a democracy is also enlightening. Political scientists tell us that Australians prefer an arms-length democracy that allows them to make a choice at election time, and then expect the politicians to set the specific agendas and take care of issues (Johnson 1987; Ward 2001). This has obvious implications for the power of political leadership to focus public understanding and action on an issue like climate change.

As we will see, the evidence over 25 years for Australia is consistent with an understanding of how information is framed and presented to the public as 'reality' with a dramatic change in what we thought we understood about climate change settling in by the mid-1990s, with the earlier understanding being gradually forgotten.

4. What Australians knew 25 years ago

The awareness of the greenhouse issue is probably greater amongst the general public in Australia than in any other country in the world

Ann Henderson-Sellers & Russell Blong, *The greenhouse effect, living in a warmer Australia*, 1989

Australia's early good knowledge of climate change was documented in a well-credentialed 1989 book that came to a startling conclusion. Following two national greenhouse effect science and public knowledge events staged in 1987 and 1988 by the national science agency the CSIRO and the federal Commission for the Future, earth scientist Ann Henderson-Sellers and her co-author Russell Blong reported on the outcomes of a two-year media and public awareness campaign. They felt able to claim that 'the awareness of the greenhouse issue is probably greater amongst the general public in Australia than in any other country in the world' (Henderson-Sellers & Blong 1989: 155).

Public knowledge was also borne out in opinion polls. A September 1988 poll reported in *The Sydney Morning Herald* began with the following headline and lead: 'Most want action over the greenhouse effect. Three-quarters of Australians are troubled by the environment-threatening greenhouse effect and believe something must be done to halt it, the latest Saulwick Herald Poll shows' (Carney 1988: 5). This was just one of hundreds of articles from that period examining science and policy on the greenhouse effect.

A questionnaire prepared by Henderson-Sellers and Blong revealed that a majority of respondents worried about the nuclear power option as a response to the greenhouse effect. The same people understood the link between greenhouse effect action and lower use of fossil fuels, and they worried about higher temperatures and rising sea levels. People confessed to a lack of scientific understanding, but wanted to know more. Perhaps most interesting in regard to the 1990s sceptic debate and related framework of uncertainty, was that a majority of respondents demanded only 50–70 per cent certainty from scientists before action was justified.

At that time, certainty was essentially available. The language in the first United Nations-sponsored Intergovernmental Panel on Climate Change (IPCC) report in 1990 was plain English and definite, and it set a communication benchmark that is commonly overlooked in research discussions of IPCC reports. Unfortunately, in the communication worlds of science and policy, the 1990 IPCC report may in hindsight be seen as a refreshing anomaly. Henderson-Sellers and Blong report that signs of mainstream scientists returning to safer conventions of scientific reporting were apparent as early as 1989. They write that at a public presentation 'Considerable surprise was expressed that scientists should be vehemently debating small differences of certainty ranging from 95–99%' (Henderson-Sellers & Blong 1989: 166).

They also found the public that they interviewed, while admittedly not the 'man-in-the-street', had a sophisticated understanding of how society interacts with such an issue. They asked whether people felt there was any attempt from any sector to deliberately confuse the scientific issue. Twenty-nine per cent of respondents thought so. The respondents, who as a group were better educated than a random poll, thought journalists and politicians were largely to blame, while scientists were seen as somewhat responsible, but not very.

Other agents of confusion nominated were multinational corporations (self-interest) and extreme environmentalists (propaganda). The authors conclude that all those surveyed, including high school students, correctly understood the scientific message, while interpreting the response of politicians and planners as ineffectual and possibly uncaring. The young people were described as seeing an unsatisfactory future, but not seeing a way to change the outlook.

Two state surveys of Australian public attitudes that were published in 1989 come from the Electricity Commission of New South Wales and the State Electricity Commission of Victoria (SECV). They are evidence that states at that time were starting to act on public knowledge with a view to containing consumer demand for coal-fired electricity. Concern about cutting down forests, the hole in the ozone layer, and the greenhouse effect were most frequently mentioned as top world environmental problems. In the NSW survey, conducted by the Roy Morgan Research Centre, 95 per cent of respondents had heard the term 'greenhouse effect' and 41 per cent knew it was warming the earth, although an almost equal number confused it with ozone layer depletion. Respondents nominated running a car, burning coal and logging forests as primary causes (along with the ozone-depleting aerosols). People also expressed themselves willing to pay more to have a large impact on emission reductions (Morgan 1989).

The December 1989 SECV survey was a small, self-selected sample in response to a discussion paper on alternative responses to 'the greenhouse challenge'. *The SEC and the greenhouse effect* found that respondents were in favour of an even stronger target for emission reduction than 20 per cent, people understood the benefits of efficiency measures, and said coal-fired electricity should not be promoted for home heating and hot water heating in preference to gas and solar. Renewable energy was supported and respondents said hidden subsidies to status quo industries should be removed. Tree-planting programs were strongly supported. Respondents even pointed out the severe conflict between wanting to attract energy-intensive industries with cheap coal-fired electricity and, on the other hand, reducing CO_2 emissions. People noted that alternative jobs could be created with clean-power industries (SECV 1989).

Evidence that people knew the risks of greenhouse gas emissions

There is ample evidence of a public discussion up to 1992 on the risks of adding to atmospheric greenhouse gases by burning fossil fuels. From greenhouse conferences and popular science books to government documents supported by a steady stream of media articles, the public was informed about risks posed by climate change and their obligations as global citizens.

Just the books published in 1989 leave no doubt about the considerable knowledge of climate change that was available to the public 20–25 years ago. Besides Henderson-Sellers's and Blong's book, another four books on the subject were published in 1989.

In their book *The greenhouse effect: a practical guide to changing climate*, the conservation campaigner Stewart Boyle and *Guardian* newspaper journalist John Ardill wrote about climate change with uncommon style and understanding of what people 'hear', or relate to, like weather analyses:

> Many of 1988's droughts and floods, heat waves and hurricanes were random events, the roll of the dice. But the dice are being weighted. In coming years they will fall hot and stormy-side uppermost more often. Hard-nosed politicians with voters to cosset, powerful vested interests to satisfy and rivals to guard against began to talk like prophets, ecologists and utopians … they began to talk of a world that is frugal and fair. (1989: 4)

Boyle and Ardill reported that, at the time, politicians appeared to have achieved a 'glimmer of visions' and that this was framed in language that spoke of solidarity, equity and accountability—in other words, an ethical framework—rather than the tyranny of the immediate. And they put this unusual political focus within the context of worldwide weather catastrophes that marked 1988:

> In 1988 the atmosphere came within one per cent certainty of proving that humanity has upset its natural balance and that it will strike back blindly and with catastrophic unpredictability. Global warming is the threat that bundles up all our woes into one problem and one solution. (Boyle & Ardill 1989: 5)

Other significant books were written by Fred Pearce, a long-time environmental correspondent for *New Scientist* magazine (Pearce 1989) and by physicist Ian Lowe, who was at the time acting director of the forward-thinking, strategic Commission for the Future and a faculty member of the Science Policy Research Centre at Griffith University. Lowe summarised for a lay audience the science and policy understanding of 1988–1989 following two groundbreaking greenhouse conferences and after encountering tremendous public interest in the subject (Lowe 1989).

The Commission for the Future, established in 1986 by then federal science minister (1983–1990) Barry Jones, was to provide a think tank environment and a public awareness forum for science and innovation developments. While the commission disappeared in the early 1990s, along with other structures from the Hawke Labor government it was influential in informing the public while it lasted.

The books from this period challenge the notion of an incremental, one-way path towards greater political and public understanding over the course of the next 20 years and up to the present. For example, Boyle and Ardill quote Mostafa Tolba, then executive director of the UN Environment Programme (UNEP) who said, 'Political leaders now accept the broad scientific consensus that human activity is altering climate and that the changes and their impacts will become more pronounced over the next few decades' (Boyle & Ardill 1989: 6).

Together with the newspaper record and other documents, these books provide a science history of events, understandings and values—at least in the English-speaking world at that time, and of Australia's climate change knowledge and response in the late 1980s.

Lowe is one of the small group of Australian scientists and researchers (with training that qualifies them to be considered expert)[1] who have consistently written and spoken about climate change and its risks to civil society, even in the face of a decade and more of increasing scorn and scepticism from powerful politicians, think tanks and opinion columnists in the media, specifically in the News Limited media and in commercial talkback radio (which scorn is still at fever pitch today).

Lowe's 1989 book *Living in the greenhouse* was followed in the mid-1990s by *Living in a hothouse*, now out of print. The first book offers the chapter and verse

---

1   Other Australian scientists/technologists from this era, who have continued to publish for a lay audience and have disputed prevailing economic ideologies driving responses to climate science during the 1990s, include Mark Diesendorf and Alan Pears. This is in addition to atmospheric scientists—CSIRO's Barrie Pittock, Willem Bouma, Graeme Pearman and Michael Raupach *inter alia*—who have spoken in many public fora and published for decades on this subject.

of what was known at the time on climate change—its risks and solutions—in notable contrast to public discussion that developed through the mid-1990s and into the present.

Delving into *Living in the greenhouse* provides the whole story of discovery from the 1890s on. The natural greenhouse effect balance of gases has been beneficial in keeping the earth warm and habitable. But:

> Our contemporary problem is that human actions since the Industrial Revolution have been changing the composition of the atmosphere … the scientific community has been concerned for several decades. By the 1950s it was suggested that the rate of burning fuels such as coal could be changing the amount of carbon dioxide in the atmosphere. (Lowe 1989: 1)

What stands out from the publications of this period is the matter-of-fact, declarative language, clearly linking atmospheric pollution and human actions. No debate there. The descriptions assume that the chemistry underpinning the science is basic and easy to understand. Lowe notes upward trends of coal burning globally (from 1.5 million tonnes annually in the 1920s to an estimated 20,000 million tonnes 60–70 years later) and the simple chemistry of burning carbon + oxygen = carbon dioxide.

As early as 1980 the Australian Academy of Science organised a conference in Canberra to review the thinking of leading scientists on the issue (Pearman 1980). Lowe says 'It was noted then that carbon dioxide levels were increasing quite rapidly and it was estimated that the pre-industrial level could be doubled by the year 2030' (Lowe 1989: 2).

The authors writing in 1989 were aware of atmospheric modelling work soon to be summarised in the 1990 IPCC report suggesting temperature increases of 1–2 degrees Celsius (°C) near the equator and 4–6 °C at higher latitudes within the 21st century under 'business as usual' scenarios. These predictions have hardly changed. (By the mid-1980s the world was experiencing a 0.5 °C average increase). Looking back, Lowe said in an interview in 2007, 'We've known for 20 years the impacts but we underestimated the speed of change; numbers have changed remarkably little. Climate change is happening a little faster.'

In the late 1980s scientists were able to confidently paint the macro effects—such as sea-level rise and changing, extreme, and unpredictable weather events— but they could not be specific about local and regional effects. This inability to predict what would be happening in particular locations was another reason for the more qualified language coming from scientists when they spoke to the public or the media in the 1990s and beyond. Unfortunately the later talk of 'probable' and 'likely', even when there was 95 per cent certainty about the likelihood of the outcome, tended to fuel debate because politicians and the media don't engage with the grey area of scientific uncertainty and tend to hear 'we don't know'.

Newspaper reports from the late 1980s indicate that scientists were signalling unambiguous confidence that the greenhouse effect was a real phenomenon caused by human activities. A review of 25 stories published in *The Sydney Morning Herald* for half of 1988–1989[2] shows that most of them quoted US scientists and referred to Australia in a global context. Potential consequences were openly discussed by government scientists and technologists. For example the following article quoted Australian scientists, with the headline 'Scientists warn of islands' peril':

> Australia may need to take in a wave of environmental refugees from coral atolls in the Pacific and Indian oceans, according to two scientists. The islands' inhabitants face being displaced by a likely rise in sea level due to the greenhouse effect, they say. The prospect was raised yesterday at the 26th Congress of International Geographical Union in Sydney by Dr Peter Roy, of the NSW Department of Mineral Resources, and Dr John Connell, of the University of Sydney. Up to about 500,000 people living on small coral islands in the two oceans could be displaced if the predictions of a one-metre rise in sea level over the next 50 years prove correct. (Quiddington 1988: 7)

Looking back in 1993, Henderson-Sellers credited the involvement of scientists and clear communication for prompting policy action. Besides being direct and to the point, early descriptions of human-caused climate change by scientists and reporters tended to address what lay audiences were likely to 'hear' from their own past experience. They used the language of risk, or current weather events, or likely impacts, such as sea-level rises. There was also an early understanding that 60 per cent or more global reduction in emissions was the necessary response and there was, therefore, a matter-of-fact assumption that the public interest required a strong response.

This assumption is reflected in the statement of then chairman of the CSIRO, Neville Wran, who told *The Australian Financial Review* in 1988 that regulation might be needed to achieve emission cuts. 'The Federal government may have to

---

2 Newspaper sampling for this analysis focused on Fairfax-owned media with one national business newspaper, *The Australian Financial Review*, and one metropolitan and regionally distributed general interest paper, *The Sydney Morning Herald*. The editorial content of these two publications over time, unlike the Murdoch press, has not been studied elsewhere. Samples were taken from years bordering IPCC reports in 1990, 1995, and 2001 as well as 1988/89. Key words like greenhouse effect, global warming and climate change yielded a pool of articles with samples taken from a consistent period at the beginning and end of the sample years.

bring in laws to control the greenhouse effect ... legislation would be required to either recognise international agreements on controlling the greenhouse effect or to regulate the phenomenon in Australia' (McKanna 1988: 4).

A senate inquiry in December 1989 also did not hedge its language and the clarity of its findings provides a useful contrast to the confused discussions we witness today. The inquiry was briefed to look at the contribution that Australian science and technology could make to combat the greenhouse effect. To do so, the Senate Standing Committee reported that it met with Pearman and Lowe; Dr John Zillman, Director of the Bureau of Meteorology; and Nelson Quinn, a senior officer from DASETT (the department of environment). The report accepted the science of the physics and chemistry, the predicted impacts for Australia, the risks, and the moral obligation as a global citizen—as shown in the following extracts from the introduction (Senate Standing Committee on Industry, Science and Technology 1989).

> The experts with whom the Committee met confirmed that there is *irrefutable scientific evidence* that the composition of the atmosphere has been, and continues to be, altered significantly by human activities [discusses ice core evidence in particular] ... The changes that are likely to occur as a result of these changes in the atmosphere cannot yet be predicted precisely. However, the scientists *predict with a high degree of confidence* that a global warming of between 1.5 and 4.5 degrees centigrade can be expected to occur by 2030. Climatic records indicate that this warming may already be happening. This phenomenon is popularly known as the greenhouse effect. (my italics)

On likely impacts the report stated:

> The sea level can be expected to rise between 0.2 and 1.6 metres, as the oceans become warmer and expand. There will be changes to the climate ... In Australia the prevailing weather patterns are expected to move south. Some areas will receive more rain but it can be expected that droughts will become more frequent in other areas, and that climatic extremes such as cyclones will occur as far south as Brisbane.

> There is a risk that if the response to this problem is delayed until the evidence of significant climatic change becomes irrefutable, it may be too late to avoid some of the more extreme changes that could occur ... Early action is essential to stop or slow some of the more extreme effects ...

The senate document shows that the political framing was moral and sought opportunity. Since our per capita emissions are large, 'we would not be in a position to seek change elsewhere unless change is implemented here'. The document said Australia should 'serve as an example' and 'develop industrial techniques and innovations'.

Indeed, the global, ethical approach is a standout feature of this time in Australia's history (McDonald 2005) briefly shared by other English-speaking countries. Ethical responsibility was seen owing both to the rest of the world and to future generations.[3]

In July 1989, Prime Minister Bob Hawke declared in his Statement on the Environment:

> The growing consensus amongst scientists is that there is a strong possibility of global warming with major climate change, and that this is linked with the levels and nature of industrial and agricultural activity. Significant climate change … would have major ramifications for human survival … (Hawke 1989: 28)

Earlier still, influential Labor powerbroker and then federal environment minister Senator Graham Richardson concurred in a *Sydney Morning Herald* interview, calling the greenhouse effect the greatest threat facing Australia and the world (Seccombe 1988).

Cost of inaction known

The evidence in these documents indicates that politicians and their advisers at that time were establishing a narrative for early intervention, as a good global citizen open to regulation for the common good. This remained a rhetorical goal at the time of Australia's participation in the United Nations Framework Convention on Climate Change (UNFCCC) 1992. Also known to policymakers at the time was the 1990 IPCC assessment of the cost of inaction or sticking with the status quo: 0.7 °C of additional warming by 2100, which is represented by the blue curve in the figure below.

The graph comes from a federal government publication released in 1992 showing how matter of fact this knowledge was at the time. The National Greenhouse Advisory Committee, chaired by ANU ecologist Henry Nix, whose main brief was to fund research, was borrowed by then federal environment minister Ros Kelly to explain climate change science to the general public. The resulting publication used plain English as it transferred the messages from the 1990 IPCC report and corrected some sceptic refrains, such as the argument that

---

3 This definite and ethical approach is highlighted in a surprising speech given by then British prime minister Margaret Thatcher to the United Nations in 1989, quoted at some length in the next chapter.

changes to the world's climate have occurred before and will again naturally—which ignores the unprecedented rate and rapidity of change caused by human activities.

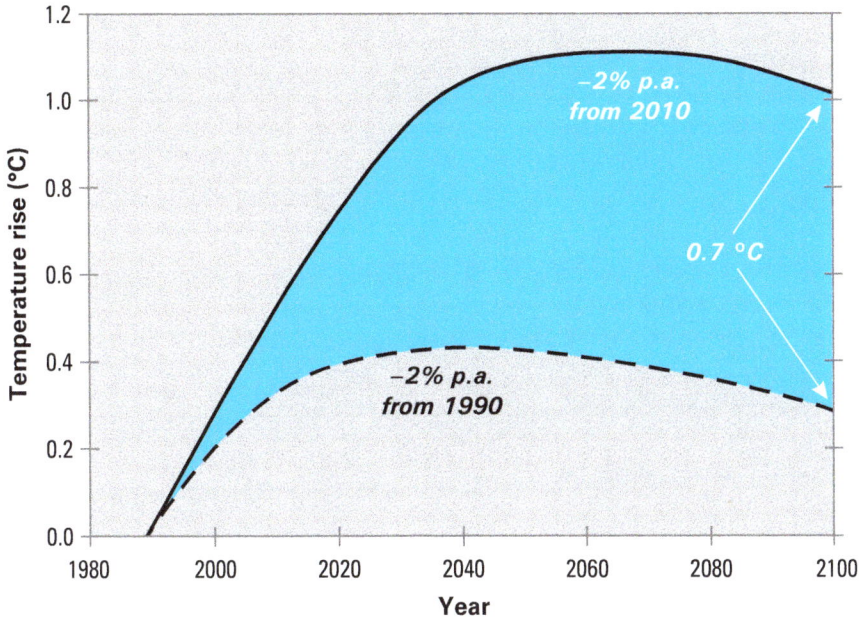

'Warming to the year 2100 is shown for two scenarios: 2% reduction per annum in greenhouse gas emissions starting in 1990, or 20 years later in 2010. The difference is the "temperature penalty" for delaying reductions.'

Source: Mitchell 1992 citing IPCC 1990: 40.

The publication noted: 'Climate change induced by the enhanced greenhouse effect represents change to the planet's climate system at a greater rate than experienced for at least 10,000 years' (Mitchell 1992: 41). (The comparison has since been amended to at least 100,000 years and, more recently, as the CO_2 level continues to rise, to 400,000 years).

Weather framed risk

The compelling climate change narrative in the late 1980s and early 1990s was often anchored by the observation that the weather was getting hotter with droughts, heatwaves, and changing rainfall patterns around the world. This was gaining public attention thanks to media stories from the period along with the popular science books from 1989.

Consider this 1988 article in the business-focused *Australian Financial Review*, aptly titled 'Government officials start to feel the climate of change':

> ... yesterday, Queensland's Water Resources Minister, Mr Don Neal, was at the forefront of the discussion. He pointed out the possible economic impact on Governments from increased flooding, more severe droughts, the effect on agricultural and pastoral industries and the need to redefine engineering design codes for roads, bridges, railways, dams and even housing ... 'There is no longer any serious doubt that climate will change more rapidly over the next 50 years than ever before in natural history'. (Massey 1988: 28)

Given 2010–2012 weather events in Queensland, this historical record becomes even more interesting. Risk messages framed the expected weather changes— with likely short and longer term impacts known at the time including the following unpleasant consequences of unchecked human greenhouses gas emissions: temperature rise (tropical cyclones); changes in average rainfall and intensity (floods); sea-level rise (landslides); droughts (wildfires); and land degradation and health consequences. New data have *refined* regional detail of risks, but the record shows the macro impacts were all known by 1988.

One could not get more 'popular' in publications than *Newsweek* or *Playboy* both of which explored the greenhouse effect in the 1980s ('Mother nature's revenge' 1987; Shears, 'The greenhouse syndrome',1980).

In US Congressional testimony, leading climate scientist James Hansen of the National Aeronautics and Space Administration (NASA) testified that he was 99 per cent certain that the warmer weather of the 1980s (along with a headline-grabbing drought) was a sign global warming had started. This reportedly galvanised legislators into considering rapid action (White 1990). Australian media reports overwhelmingly heeded this international response as well as active domestic science communication. The target audience was the Australian mainstream and the message was framed as risk to everyone. In June 1988, a report from Paul Sheehan in Washington D.C. published in *The Sydney Morning Herald*, states:

> Scientists have warned about the 'greenhouse effect' for years. Now it is no longer a scientific nightmare; it has arrived. ... The 'greenhouse effect' is the term given to describe the gradual heating of the Earth's atmosphere caused by the increasing production of fossil fuels and pollutants. (Sheehan 1988)

Risk management was also on the government's agenda, as shown in a 1987 federal environment department (DASETT) briefing minute to the CSIRO Division of Atmospheric Research. It said 'risk management was necessary' and that 'action needs to be taken now' ('Climate change due to the greenhouse effect' 1987: item vi). The document speaks of more extreme events and erratic weather. It urges action despite scientific uncertainties on specifics, but understands the risks posed by human activity creating the greenhouse effect. A similar understanding is evident in numerous other government documents published between 1987 and 1991.

The important Toronto conference 'The Changing Atmosphere', which was held in 1988, saw the Canadian Government host some 300 scientists, politicians, and economists from 48 countries. The conference coincided with severe drought and high temperatures in North America. The events of the conference, and its dramatic final statement of urgency and call for emission reduction targets, were widely reported in the Australian and international media. *The Sydney Morning Herald* reported on 2 July 1988 that the international call to action was attempting 'to reverse the trend towards a hotter, drier, carcinogenic world before the pace of environmental deterioration accelerated beyond man's ability to stop it' (Benesh 1988). This is one of many articles from the period that accepted the scientific risk assessment as fact *before* the first IPCC report.

Toronto helped focus world policy attention, reaching for an action model that seemed to have worked to curb ozone pollution—set targets internationally and let governments work out the policies to meet them (Weart 2004). The conference's experts proposed reduction of CO_2 emissions to 1988 levels by 2000 and a further 20 per cent reduction by 2005, a formula picked up by Australia and other countries at the time.

1988: Coming to grips with a terrifying global experiment

The Toronto conference statement made it clear that climate change would affect everyone. It called greenhouse gas atmospheric pollution an 'uncontrolled, globally pervasive experiment whose ultimate consequences could be second only to nuclear war'. World governments were urged to swiftly develop emission reduction targets (*The changing atmosphere: implications for global security*, 1988).

Toronto paved the way to the United Nations and World Meteorological Organisation IPCC structure, which was also established in 1988.

Hundreds of climate scientists, economists and national policy representatives were asked to review the global research on causes, impacts and potential responses (three working groups) and report to the second world climate conference in 1990 in Geneva and every five years thereafter. Science journalist Fred Pearce noted the irony that the IPCC concept was promoted during the US administration of President Ronald Reagan as an effort to contain or dilute noisy government scientists talking about dangerous climatic changes as they did at their groundbreaking 1985 UN-sponsored conference in Villach, Austria.

> The purpose of the IPCC was to put scientists back in the cages they had briefly escaped from at Villach, and to this day the IPCC's members are government nominees. But it was too late. The story of global warming — and what scientists really felt about it — was out. (Pearce 2005: 53)

The scientists who gathered at the 1985 Villach conference made significant risk pronouncements linking anthropogenic increases in greenhouse gases with global warming and climate change. A consensus emerged within the climate science community that 'we have a problem' (Bolin et al. 1986; Pearman 1988).

At the same time, the multi-decade struggle to understand and ban ozone-depleting substances, another major man-made risk to planetary health, was drawing to a conclusion. In 1987, the international Montreal Protocol banned most of the traditional chemicals that had been used in refrigerants and powered aerosol sprays. This seemed to demonstrate that countries *could* come together for the good of their living environment, based on scientific advice. (Unfortunately, the two issues—the hole in the ozone layer and greenhouse gas pollution—also became confused in the public mind).

It may surprise those who think the scientific evidence was thin at that time, but the Villach statement in 1985 reflected conclusions drawn from much previous research, ranging from ice-core evidence to analysis of atmospheric chemicals, experiments with general circulation models, and observations on heat and sea-level rise (Lowe 1989).

By 1988 politicians were becoming more involved. Graeme Pearman called the evidence facing the Toronto global conference of concerned scientists and bureaucrats a 'clarion call to politicians to take action' (Lowe 1989: 4). Pearman was an author of the 1990 IPCC science working group report.

A brief open door of opportunity

The 1990s political changes that would bury Australia's early climate change response strategy make the events of the late 1980s all the more remarkable in hindsight. Early public awareness in Australia probably reached its apex as the CSIRO, along with the federal government's Commission for the Future, developed two national greenhouse conferences featuring Australian climate change science of international standing. The conferences garnered widespread media and community attention.

An important element of the 1980s history of climate change understanding in Australia and overseas was the leadership by scientists and their effective interaction with media and policymakers as well as the organisation of conferences and major public events. European environmental journalist Fiona Harvey told the (US) Society of Environmental Journalists in 2006: 'When people first heard about global warming, it wasn't from politicians, it was from scientists through the media. So we got the scientific view before any politics got attached to it' (quoted by Thacker 2006).

In the United States, media research has suggested the same phenomenon: media stories peaked between 1988 and the early 1990s, and scientists were a primary source of information during that early period. Environmental journalists recalled that media coverage picked up again around 1997, when the Kyoto Protocol was under discussion (Wilson 2000).

The December 1987 CSIRO 'Greenhouse '87' conference was primarily a meeting of scientific experts who were given a baseline that climate change would happen and were asked to analyse the most likely impacts and scenarios (Pearman 1988; Lowe 1989). 'The conference attracted considerable media interest, with the main emphasis, perhaps predictably, being on the worst cases of gloom and doom: coastal land possibly flooded, agricultural areas possibly turned to desert, cyclones possibly moving further south, and so on' (Lowe 1989: 4). The media appetite that emerged for the most dramatic possibilities may have encouraged scientists themselves to draw back from such black and white predictions and the definitive language that characterised the first (1990) IPCC report.

A year later, another CSIRO event was organised with the federal government's Commission for the Future. 'Greenhouse '88' has been characterised as 'extraordinarily ambitious' (Lowe 1989: 5) with conferences in all capital cities plus Cairns. Local meetings were held to establish planning committees for future action. The committees were composed of generally well-informed individuals committed to improving the level of community awareness.

US atmospheric scientist Stephen Schneider gave the keynote address at 'Greenhouse '88', saying scientific consensus and evidence was sufficient to take action (Lowe 1989: 6). Other speakers included: Barry Jones as science minister; then Victorian minister for education (and soon to become premier) Joan Kirner;

and commission chairman, broadcaster and writer Phillip Adams—and the documentary record shows many organisations, public and private, were also involved.[4]

Schneider, who died in 2010, played a significant role during much of the early history of climate change communication. It started with his 1976 book *The Genesis Strategy* in which he famously predicted that in 2000 the effects of human activities will emerge from the background noise of natural climate variation (Schneider 1976: 228), which proved to be close to the mark. During the 1990s, however, he also argued persuasively to diffuse the language of the IPCC to emphasise the uncertainties. Melbourne *Age* journalist Geoff Strong, who reported on climate change during these years, observed that raising the profile of uncertainties unwittingly played into the hands of denialists and naysayers in the decades to follow (which is covered in more detail in the following chapters).

Local discussions following 'Greenhouse '88' continued for two days and organisers counted 8,000 people involved, claiming it was the largest conference ever held on an environmental issue. Being widely reported, it garnered political attention across the spectrum. Then environment minister and Labor powerbroker Graham Richardson, for example, was one who spoke out at the time.

His role and sincerity on the climate change issue is still debated, with many contemporary observers guessing that his interest was purely political and involved 'counting the numbers' of potential votes for the next election, which, in itself, would stand as testimony to the strength of public awareness. Regardless of motive, his utterances in the media record reveal a solid understanding of the magnitude of the risk. Ian Lowe told me in an interview, 'Richo at first was just a pragmatic political fixer but when (Greens leader) Bob Brown took him to the Tasmanian forests, he became a convert'.

Through his involvement with the Commission for the Future, Lowe had firsthand experience with the contemporary public interest and wrote his 1989 book on climate change in part to satisfy that interest. He testifies to a great deal of media coverage:

> The mass media took up the question of possible climate change with great enthusiasm. The *Age* published a 4-page supplement in association with the Commission. TV programs *Quantum* and *Beyond 2000* made

---

4   In his 1989 book *Are We Entering The Greenhouse Century?*, Schneider recalled his invitation to speak at 'Greenhouse '88' by Philip Noyce, deputy director of the Commission for the Future, and the intense round of media interviews throughout the country that were organised preceding his speech. He found (to his relief) that both the media and general public were well informed and asked good questions, thus providing further evidence of Australia's good public knowledge at this time.

special editions, a special Sunday Conference was devoted to the issue and it seemed to be on every radio station. Suddenly it seemed that everyone knew about climate change. Radio stations aimed at the youth market were particularly keen to take up the issue, reflecting their awareness of the great concern of young people about environmental issues (Lowe 1989: 6).

Australia's early emission reduction targets 1989–1990

Following these national events and in response to the Toronto targets and international calls for action, by 1989 Australian state governments released initial greenhouse response strategies focused on energy conservation and substitution for coal. The need to retain native vegetation was acknowledged (and in 1990 would be supported by the Hawke government kicking off the decades of Landcare and tree planting).[5] Local government looked at specific impacts, and schools and voluntary groups got informed and involved. To help them out, the Commission for the Future released the *Personal action guide for the Earth* that again shows the early understanding of the consequences of the greenhouse effect and that the science findings have hardly changed in a quarter of a century (Commission for the Future 1989: 5).

In October 1990 Australia formally adopted the Toronto targets by setting a so-called 'interim planning target' of stabilising greenhouse gas emission at 1988 levels by 2000, and reducing them by 20 per cent from that level by 2005 (Commonwealth 1990).

Energy efficiency headed the response strategies talked about at that time, framed as a way to cut emissions significantly and as a win-win with cost savings. Early 1990s Commonwealth fact sheets for the public promote efficiency measures and fuel substitution for heating, lighting, and transport (e.g., use of insulation, passive solar, fluorescent light bulbs and solar hot water, heat recycling, conversion to gas).

They promised that these strategies could yield an 18.8 per cent cut in carbon dioxide emissions over time, in line with the interim planning target of cutting at least 20 per cent. At the same time, consumers and businesses could save millions in energy bills (*Climate Change Program*, Commonwealth n.d.). As we shall see, this efficiency approach—called demand management—eventually

---

5   The Landcare movement officially began in 1989 when the Australian Government formed the 'Decade of Landcare Plan' in an effort to protect, restore and sustainably manage Australia's natural environment and productivity.

stumbled in the face of the growth and 'sell more electricity' lobby as state-owned energy suppliers were corporatised, and/or privatised, and found themselves competing in a national market.

In so many ways back in 1989, with the weather giving warning signs, and media plus political leadership on board, the documentary evidence indicated that appropriate action would follow step by step. And so it did—on paper. Yet, within 10 years these messages had been reframed into a hazy 'scientific debate' characterised by uncertainty, which confused the public and blocked action.

Even at the early and active stage, some commentators were cautioning about the length of time it might take for effective global action. Lowe, for example, harked back to the slow trajectory of action on the ozone hole—between 1974, when discovery of damage by chlorofluorocarbons was made, and 1987 (13 years later) when degeneration of the ozone layer became measurable. In the event, the climate change story is taking longer to sink in, subjected to comparable forces of denial and manufactured uncertainty.

The ozone story provides a parallel about communication and issues within the scientific community itself regarding empirical evidence. 'For 13 years those who wanted to do nothing could stall by saying the evidence is not good enough, only *measurement* is evidence that warrants action,' Lowe said in an interview for this story.

For communication scholars, another important lesson when comparing the two phenomena under similar sceptic and industry attack is that the ozone hole risk enjoyed the advantages of being easily described with a metaphor (a hole in the earth's protective shield), it denoted an urgent crisis (skin cancer risk) and there was a comparatively simple solution to hand (Ungar 2000).

Emission reduction stalls at reality of energy markets

Despite the understanding of urgency to act, it was recognised early that somewhat mitigating the risks would be a major undertaking, let alone reducing emissions by 60 per cent as the IPCC was suggesting might be necessary. The 1990 IPCC report on response strategies warned that, 'Achieving a 20 per cent reduction from current emission levels would require major changes in global energy markets, plans and infrastructure and intervention by governments' (Bernthal 1990: 66–67).

A 1989 report from the Australian Prime Minister's Science Council noted that reaching this same target would call for 'deep-reaching and pervasive energy-restructuring at considerable cost and substantial government intervention' (Kolm & Walker 1989: 147) foreshadowing major barriers that would be erected to action in the next decade, namely the cost barrier, the power of the status quo and the ideological dislike of regulation.

Inefficient Australian industry in export environment: change the story, not the industry

In the late 1980s the media conducted an open discussion about the state of Australian fossil fuel-based industries. Top of the agenda for a different energy production profile was conservation and efficiency and fuel conversion (to gas), renewables, along with related new job creation. The following article from *The Australian Financial Review* discusses the federal government's push for energy efficiency, and says Australian industry is extremely wasteful in its use of energy:

> ignoring even simple energy-saving projects that would have a payback time of less than one or two years. The study found that energy consumption and greenhouse gas emissions could be cut by 15 per cent using easily identifiable energy savings that would actually make rather than cost money.
>
> …
>
> The bigger picture is that the industrial sector contributes 36 per cent of all Australian emissions of carbon dioxide, the main gas associated with the greenhouse effect. A national target of a 20 per cent cut in emissions by the year 2005 is thought likely to be adopted in the near future. If this is to be achieved, then industry is going to have to show it is taking the problem far more seriously. Otherwise it will only be inviting the Government to force it to take action. (Roberts 1989: 17)

In other words, it was recognised in 1989 (and in the business media) that halving emissions from an inefficient industry sector would get Australia a long way towards its target of 20 per cent cut in emissions. There is also early evidence of the conflict created by Australia's economic direction towards expanding minerals extraction, and specifically a focus on coal, that would come to dominate in the 1990s. The following *The Australian Financial Review* article made public the conflict between emission reduction and the push towards 'quarry Australia':

> The reasons for Australia's awkward position on global preventative responses to the greenhouse effect are simple. Ours is a carbon intensive

economy. We are the biggest exporters of coal in the world. Electricity generation has been growing by 6 per cent per annum over the past two decades, to account for 44 per cent of carbon dioxide emissions. Ninety-five per cent of electricity produced in Australia is generated by the burning of fossil fuels such as coal and gas.

Per capita, Australians are the fifth highest greenhouse polluters in the world …

Moreover, to trade out of its foreign debt burden, Australia hopes to do much more processing of raw materials in the 1990s. Like the wave of aluminium smelting investment of the early 1980s, this would be energy intensive—and thus greenhouse intensive—stuff. (Stutchbury 1990: 16)

Fast forward 23 years and the documentary record indicates that *The Australian Financial Review* since the mid-1990s has moved a long way from its editorial position in the 1980s and early 1990s as a critic of business practices when necessary for an informed public.[6] Early *Financial Review* articles, such as the above, illustrate a more neutral and sometimes critical stance on Australia's energy-intensive industries. In those early years the publication also reported the climate science story much as other mainstream papers did. That changed to a strongly partisan pro-industry position by the mid-1990s.

The 1990 Stutchbury article foreshadowed that Australia's minerals and smelting sector would drive an argument that Australia had a special case not to make energy sector changes or significantly lower greenhouse gases—because it was an inefficient sector that was sensitive to increases in the price of electricity if it wanted to remain competitive on the world stage.

Instead of including that sector in efficiency drives and lower electricity use, a succession of Australian policymakers and industry lobbyists during the 1990s decided to keep the energy production and use status quo as it was (Pearse 2007), and instead change the framing of the public narrative to suit that goal. (To the extent that energy-intensive industries still operate inefficiently, this may partly explain the hardline resistance to carbon pricing by their allies in the federal government under Tony Abbott).

Also revealing where the country was headed was a November 1989 report, again in the *Financial Review*, that the Treasury under Paul Keating—in an internal argument about emission reduction targets—was advocating the position that Australia could increase its pollution as a specialist energy user.

---

6   The reporter of the above article, Michael Stutchbury, after a stint as economics editor at the Murdoch-owned *Australian*, returned recently to *The Australian Financial Review* as editor-in-chief and, in a 2013 ABC panel program, defended the position of the coal industry in national affairs. The political website *Crikey* has called the *Financial Review* in its current incarnation an elite outlet for big business messages (Dyer & Keane 2013).

It was reported that:

> Senator Richardson wanted to set a 20 per cent reduction target by the year 2005 but was defeated by the intervention of the Treasurer, Mr Keating. The Treasurer convinced Cabinet that Australia should instead promote itself as an energy-efficient industrial centre. The argument was that while pollution might increase in Australia, there would be an overall reduction worldwide. (Dunn 1989: 8)

Industry opposition, sceptics on horizon

The first rumblings of opposition from the mining sector started at this time too. An early example comes from a scientific presentation at a Brisbane seminar organised by the Queensland Government in 1989. Lowe reports that Griffith University scientist Roger Braddock presented a cautious paper reflecting scientific uncertainty about the state of knowledge on interaction of atmosphere and oceans, that is: how, and how much, carbon dioxide from the atmosphere was being stored in the oceans.

> It was only too depressingly predictable that his paper would be misrepresented; the President of the Queensland Chamber of Mines wrote letters to politicians and various publications, claiming that Dr Braddock's cautious approach proved concern about climate change was unjustified hysteria from wild-eyed extremists. (Lowe 1989: 11)

Sceptic scientists were on the horizon by December 1991 when a prominent representative—Robert Balling, a climatologist from Arizona State University—was invited to The Australian National University (ANU). It was reported in the federal government's newsletter, that he called the IPCC report 'scare mongering' and said there was no evidence of a hotter planet. The visit was sponsored by the Tasman Institute, a free market think tank. Balling's visit was one of several by US sceptics who were brought to Australia in the early 1990s by think tanks, but also supported by the CSIRO and universities, in the name of free enquiry and debate (Department of Primary Industries and Energy 1992).

Sceptic and atmospheric physicist Richard Lindzen, who had some of the most relevant scientific credentials of the well-known critics, was invited to Australia by the CSIRO and addressed the National Press Club in June 1992. He is quoted in an industry conference paper saying 'most scientists in the field do not agree the case for action has been demonstrated' (Daley 1992: 3).

Still, at that time, then federal science minister Barry Jones recalls, 'I didn't have the Minister for Minerals and Energy shooting me down (and) at that time there weren't the hardball lobbyists.' He also points out, however, that even during that early period the scientists with whom he spoke did not have a unified view on the human contribution to climate change and that both the head of the CSIRO Division of Atmospheric Research, Brian Tucker, and the Bureau of Meteorology's head John Zillman, while in the minority, were personally more sceptical and also influential because of their positions. Both of these men, perhaps due to their disciplinary backgrounds, (more on that later), were more liable to express reservations about climate system modelling that could not be measured on-ground.

It is noteworthy though that, in his professional capacity, Tucker had edited a monograph on climate change science for the Australian Academy of Science as early as 1981. In June 1986 he would make a presentation on behalf of the division to the Australian Environment Council (AEC) that helped galvanise the government into more research funding and also into communication activities, as recorded in a 1987 departmental minute to the division (*Climate change due to the greenhouse effect* 1987).

According to federal politician Bob Chynoweth, the then parliamentary representative on the division's advisory board, who reported back to Jones during the mid-1980s, Tucker and his scientific colleagues also knew how to effectively operate in the policy environment: 'The real work was done by short-circuiting the bureaucracy and going straight to the minister. That's how you got things done. Lobbyists go straight to the minister,' he told me in an interview about how science and politics used to interact.

Atmospheric scientists knew the communication game

As one of Australia's lead scientists in the new field of climate change science, Graeme Pearman approached his role in much the same way. He appreciated the integrated nature of sustainability research and communicated widely. He said his strength was to get to know and engage directly with politicians, bureaucrats, community groups like Rotary, and particularly the media, and that he saw this as a serious opportunity to benefit the taxpayer's investment in the research.

Barrie Pittock, Pearman's colleague at the CSIRO Division of Atmospheric Research, wrote numerous articles during the 1980s and 1990s for professional and lay publications focusing on the state of scientific knowledge and risk, and he tackled the still uncertain issue of regional and local impacts on Australia.

The following summary comes from a paper in *Australian Forestry*, one of dozens of articles, book chapters, and conference speeches Pittock produced from 1980 to the late 1990s:

> The atmosphere beyond the year 2000 will be different from any experienced since before the last glaciation, more than 100,000 years ago. This will profoundly affect forestry locally and globally. Large percentage increases in carbon dioxide and other greenhouse gases will cause temperatures at a given location to be far higher than any in human history, and [also cause] large local changes in rainfall and humidity. These will greatly affect tree growth, species composition in natural forests, and fire frequencies. (Pittock 1987b: abstract)

A 1991 paper, co-authored by Pittock on climate change scenarios for Australia and New Zealand by 2010 and 2050, is notable also for how long the publisher had been around—the journal *Climate Change* was by then in its 18th issue. In a 1987 presentation to a lay audience at the Peace Research Centre, Pittock did not hedge his words: 'The greenhouse effect throws into question the whole global trend towards increasing population, and industrialisation based on greater energy use' (Pittock 1987a). At the time, such plain speaking and policy comment was not unusual for scientists. These public interactions became more unusual by the mid-1990s, as more scientists employed a language of uncertainty or stopped public discussions altogether.

Politicians: Many did not believe impacts *could* happen

In the late 1980s, political leaders (Jones, Hawke and Richardson) publicly interacted with the CSIRO scientists and division advisory boards. From that advisory board, Bob Chynoweth personally briefed the prime minister, according to a Hawke speech to the division on 19 March 1990 (Hawke 1990).

Chynoweth himself gave an extraordinary speech to federal parliament in October 1987, in which he laid out in clear language some of the scientific scenarios of likely impacts of a warmer world, 'a huge greenhouse' (Chynoweth 1987). Direct human impacts, he said, would include increased incidence of skin cancer and eye disease, and immune system depression and disease related to increased ultraviolet radiation. Collapse of ocean ecosystems came high on the list.

'We must now accept the very chilling announcement that mankind is actually fouling its own nest. For the first time the life habits of one of earth's inhabitants is upsetting the very balance of all life on the planet'. He reported to federal parliament more than 25 years ago the scientific prognoses of what would happen when temperatures climb 2–3°C: 'Rainfall will increase by up to 50 per cent in summer and there will be a decrease of 20 per cent in winter', and there will be more cyclones and expected sea-level rises (Chynoweth 1987).

Yet Chynoweth acknowledged that despite the open exchange between scientists and politicians during this period, most of his colleagues did not speak of or voice concern about climate change. He said that many people just did not believe the sea level could rise. Here is further evidence that leadership commitment to the issue made the difference in the policy arena in the face of beliefs and values that rejected the science findings. Such matters of belief were reinforced as the 1990s progressed and leadership changed.

'Special interest' tag for environmental lobbyists

The exchange became more adversarial as green groups gradually did more of the talking on climate change action and the focus shifted from science to the political arena in the lead-up to the Kyoto Protocol negotiations. The shift to non-government lobby groups encouraged the political and media perception and narrative that this was a 'special interest,' not a mainstream issue, championed by people who wanted to harm the Australian economy. This frame was possible because, by the mid-1990s, the domestic narrative had shifted to an almost exclusive discussion of economics and costs, with far-reaching consequences, as we see in the following chapters.

The economy was also on Brian Tucker's mind after he quit the CSIRO Division of Atmospheric Research (outgoing as chief in 1992, to be succeeded by Graeme Pearman). Thereafter he aired his sceptical views with publications through the free market Institute of Public Affairs (IPA). There he gave different policy suggestions, criticising emission reduction targets, and other responses meant to lower risk. As he viewed it, these threatened to severely compromise the national economy. He suggested 'planned adaptation' to any climate change would be the most sensible policy (Tucker 1994: 1). His involvement with the IPA think tank, along with other sceptic scientists like geology professor Bob Carter, added credibility to the IPA's attacks on atmospheric scientists and their modelling tools.

Early call to action was non-partisan

It is hard to recall from the current, bitterly divided political stage, but climate change action was non-partisan during the late 1980s and early 1990s. Far from the political split on the subject that started during the Coalition government under John Howard in the later 1990s, early policies and leadership rhetoric of both major political parties were publicly committed to taking decisive action on climate change mitigation.

Thus in 1991, the then chairman of the Australian and New Zealand Environment and Conservation Council (ANZECC) was Bill Wood, a Labor politician from the Australian Capital Territory. He wrote in a foreword to a report on the response programs proposed by the states starting in 1988, that ANZECC had noted: 'that the Panel on Climate Change (IPCC) "calculated with confidence" that emissions of carbon dioxide from human activities would have to be reduced by 60% to stabilise its concentration and that other gases would need to be reduced between 15% and 85%' (ANZECC 1991: i).

New South Wales, for one state example, was looking at mandatory insulation of homes, government leadership on energy efficiency and investment in alternative technologies, and reviews leading to new restrictions on clearing native vegetation—all under then Liberal premier Nick Greiner ('New South Wales in climate strategy' 1989).

The bipartisan consensus before 1991 was influential in allowing response measures and public understanding to proceed as far as they did. This consensus contrasts strongly with the public doubts and confusion regarding human agency in climate change that were led by Coalition politicians after 1996, and the ideological division already evident in internal bureaucratic debate under the Keating Labor government after 1991.

Will to action turns to wishful thinking on reducing emissions, later

Reflecting the good public and political understanding of the late 1980s, Pearman said in 2009: 'In the late 1980s we still had a chance to stop emissions at 350 ppm'. The figure of 350 ppm (parts per million) of CO_2 in the atmosphere was a scientific benchmark for a concentration where negative impacts are still considered reversible over the course of a century. (We might remember that stable, pre-industrial levels were around 250 ppm).

Some 20 years later, in 2008–2010, influential government economic consultants and advisers (e.g., the Garnaut Climate Change Review, emission trading scheme (ETS) modelling) assumed a course of 'stopping' (combined) greenhouse gas emissions at 550 ppm in the next decades (Garnaut 2008). The 550 ppm figure is more than double the pre-industrial level—with CO_2 emissions alone at about 400 ppm in 2013.

The relaxed 30–40 year timelines for emission reduction in these recommendations show a lack of appreciation of the long-term effect of accumulating gases. Instead the assumption is that, through some undescribed levers, gas levels and related temperatures can be eventually dialled down as convenient to the needs of national economies (Glikson 2008a). Such thinking is consistent with contemporary Australian Government and business dedication to maintain status quo fossil fuel use and coal exports with no end in sight.

The reality is that there is no option to dial down emissions and early deep cuts have major benefits, as explained for example in the 1995 IPCC science assessment. Any hope to stabilise greenhouse gas concentrations is governed more by the accumulated amount of emissions rather than by how those emissions change over time. Defending higher emissions, or very small cuts (such as a proposed five per cent cut) guarantees the need to offset by cutting more deeply in the future to have a hope of a stabilising even at a suggested 450 ppm or 550 ppm of combined gases (IPCC 1995: 3).

This is not a recent revelation. In 1989 it was already understood that a 'business as usual' approach would set the atmosphere on track for 450–550 ppm of combined gases or CO_2-equivalent (CO_2-e equals CO_2 plus methane plus nitric oxides) within the 21st century—a level of greenhouse gases that atmospheric scientists then and since have said could well lead to dangerous climate tipping points where 'feedback loops' cannot be stopped—including massive methane emissions escaping the Arctic tundra as it melts.

Atmospheric scientists set 450 ppm of CO_2-e in the atmosphere as the outer limit where accumulated emissions lead to about 1 °C warming—a scenario where change is still reversible. That means, with appropriate emission reductions, the excess CO_2 in the atmosphere could be captured within a century (IPCC 1995: 6, 15–16).

This understanding appeared lost as the climate change narrative reported in the mass media became framed by people other than specialist climate scientists, with a preponderance of economists providing media commentary. Canberra technology writer Ben Sandilands was one of the commentators in December 2008 who said the federal Labor Government's emission reduction target of five per cent of 2000 levels, based on the Garnaut review, and linked as an objective to the carbon price, did not reflect the true risks. Sandilands blamed a popular discourse that is 'scientifically illiterate' and a media 'which is too lazy to inform itself about the realities' (Sandilands 2008).

From one reporter's summary of 1990 knowledge (or, all the solutions we are considering now were known then)

The following points, extracted from a lengthy article by *Sydney Morning Herald* reporter Paul Cleary, show the extent of knowledge in 1990 reported in the media, in this case with the headline 'It's the end of the world as we know it'. This article also shows the beginning of government and industry economic modelling on cost that came to dominate the discussion in later years.

- Australia's economy is carbon intensive.
- Our output of greenhouse gases is rising at double the world average and our per capita emissions are among the highest.
- The federal government (under Hawke) wrote Cleary, 'has quite clearly embraced the concept of global warming and is keen to put in place a range of policies'.
- The first IPCC report (1990) 'provides virtually irrefutable evidence of global warming'.
- The world was heading toward a climate convention (to become the Framework Convention on Climate Change, 1992), which in turn, should lead to binding emission control protocols.
- It was thought at the time that there would be general agreement on cutting greenhouse gas emissions; the target was the Toronto goal of 20% below 1988 levels by 2005, which should be adopted by governments. The Toronto meeting of scientists and governments had agreed that significant global warming was a near certainty.
- A carbon tax on wealthy nations was seriously being considered.
- The 1988 Toronto conference coincided with a severe drought in North America and elsewhere, which ignited the media's attention.
- The government was being urged at the time to become a 'fast follower' of technological opportunities for business development related to lowering emissions.
- Substantial government 'intervention' in the economy would be required.
- Cleary accepted that 'There is little doubt that the cost of achieving such a target, both in terms of resources and standard of living, will be huge'.

Examples given by Cleary show that Australian resource industries were starting to do their own figures; e.g., coalminer CRA was warning that cutting emission by 20 per cent would hike power charges by 40 per cent; raise car prices by 25 per cent and petrol by 120 per cent—much later shown to be serious overstatements or ambit figures—e.g., by the 2006 Stern Report and the 2007 IPCC report.

- Policy responses considered included 'ironing out inefficiencies that had long been a way of life' such as:
  - state government subsidies for electric power generation that kept prices low
  - state electricity authorities should stop increasing capacity and focus on helping consumers conserve (demand management).
- Other detailed proposals worked out how much CO_2 emissions could be avoided if solar hot water were promoted to a reasonable level—8 megatonnes (MT) a year. There are similar figures for energy efficiency of appliances and refrigerators; switching to natural gas; retrofitting homes and calling for energy-efficient design of new homes as part of the building code; as well as developing energy audits—it was all there and could be achieved within 15 years (i.e., by 2005), saving about half of the 40 MT of CO_2 then emitted by households annually.
- This article does not mention fuel efficiencies and the auto industry, but those were other areas discussed at the time where efficiencies could be made, and involved federal rather than state government regulations.
- Energy-intensive industries, such as aluminium, could make process adjustments to save on electricity and low-energy intensive industries could make savings by redesigning new buildings and retrofitting old ones e.g., estimates that aluminium could cut its emissions by one third (32 MT) annually by changing process from electrolysis to direct reduction.
- Cogeneration (electricity) with natural gas could cut emissions by 10 per cent or 25 MT.
- Some cleaner coal burning options at the time, such as gasification, could achieve savings up to 25 per cent or 50 MT.
- One easy, positive outcome would be the elimination of another greenhouse gas — chloroflurocarbon emissions (18 per cent of the total)—by 1995, thanks to the global treaty to ban these gases to protect the shielding ozone layer.

Taken together, the options documented by Cleary posed a challenge to the status quo, but not a 'freeze in the dark' proposition (Cleary 1990). As government analyses commissioned at the time pointed out (e.g., Greene 1990a) there were plenty of dollar savings and job creation possibilities to make it a potential 'win-win' scenario. In the event, almost none were put into effect.

Leadership

As in Australia, international leadership was evident in the late 1980s, at least at the rhetorical level. Robert M. White, then president of the US National Academy of Engineering, wrote an extensive 1990 report about the scientific and political climate change understanding at the time (accepting human agency). He advised that governments were rushing to outdo each other on advocating action to stabilise the global climate: 'Soviet President Mikhail Gorbachev, President George (H.W.) Bush, Prime Minister Margaret Thatcher and French President Francois Mitterrand, share similar views on the climate-warming issue' (White 1990: 18).

In fact, reportedly briefed by senior scientists and advisers, Thatcher made an extraordinary speech to the United Nations in November 1989, which is worth quoting at some length because it is solid evidence of the early knowledge available to policymakers, which would not be considered outdated 20 or 25 years later. Addressing both secular and religious audiences she said:

> What we are now doing to the world, by degrading the land surfaces, by polluting the waters and by adding greenhouse gases to the air at an unprecedented rate—all this is new in the experience of the earth. It is mankind and his activities that are changing the environment of our planet in damaging and dangerous ways.

> The result is that change in future is likely to be more fundamental and more widespread than anything we have known hitherto. Change to the sea around us, change to the atmosphere above, leading in turn to change in the world's climate, which could alter the way we live in the most fundamental way of all. That prospect is a new factor in human affairs. It is comparable in its implications to the discovery of how to split the atom. Indeed, its results could be even more far-reaching.

> The evidence is there. The damage is being done. What do we, the international community, do about it? ... The environmental challenge that confronts the whole world demands an equivalent response from the whole world. Every country will be affected and no one can opt out. Those countries who are industrialised must contribute more to help those who are not.

> Reason is humanity's special gift. It allows us to understand the structure of the nucleus. It enables us to explore the heavens. It helps us to conquer disease. Now we must use our reason to find a way in which

we can live with nature, and not dominate nature. We need our reason to teach us today that we are not—that we must not try to be—the lords of all we survey.

We are not the lords, we are the Lord's creatures, the trustees of this planet, charged today with preserving life itself—preserving life with all its mystery and all its wonder (quoted in Monbiot 2005).

Who advised Thatcher to such an ethical defence of the Earth? One contention is that it was James Lovelock, an independent scholar scientist, therefore not tainted as a 'government scientist' in free market eyes (Flannery 2005: 246). Others have credited (Sir) John Houghton, lead author of the first IPCC scientific assessments and a leading UK atmospheric scientist, then director general of the UK Meteorological Office.

Even in this global context, with world leaders offering such stirring speeches, some still felt able to claim that Australia in the late 1980s was a world leader in public awareness of anthropogenic climate change. The question then becomes, how could such definite knowledge change dramatically to a sceptical debate as it did, and under what influences?

5. Australians persuaded to doubt what they knew

It wasn't raining when Noah built the ark, but at least he listened.

Anonymous, In *The greenhouse effect, science and policy in the Northern Territory*, Moffat 1992

What happened in the 1990s? Most dramatically, the fossil fuel and allied industries got into gear. The momentum to support and expand the existing fossil fuel economy was boosted by neo-liberal think tanks and insistent sceptics, in sympathy with free market economic ideology. They mounted a potent and high-level lobbying campaign aimed at federal politicians. Coal, oil, natural gas and other extractive industries, along with other multinational corporations, such as the energy-intensive aluminium smelting industry, got organised and exerted considerable influence on government, particularly after 1995 (Hamilton 2001; Pearse 2007).

This was made easier by a revolving door of policymakers and economic advisers switching in and out of senior government and business group positions. They formed a like-minded elite network directing Australia's response to the science after 1992. A picture emerges of how the focus on growth and resource extraction industries—long-standing drivers of Australian politics and the economy—undid the early good public understanding.

But in 1990 this was still not apparent. Indeed, 1990 was a highpoint for environmental politics in Australia, with federal Labor looking forward to an election supported by environmental votes, the release of the first Intergovernmental Panel on Climate Change (IPCC) report on climate change, and federal 'interim planning targets' for controlling CO_2 emissions put in place, based on the global Toronto target of 1988.

The recognition that global warming/climate change science required a policy response coincided with the Labor government under Bob Hawke sponsoring roundtable policymaking under the banner of ecologically sustainable development (ESD) in 1990. This unusual effort in democratic decision-making brought environmental, business, government, and labour leaders to the same table. The task was to determine more sustainable economic strategies and include environmental costs in the analysis. A joint taskforce was then asked to recommend how to curb greenhouse gas emissions from the energy sector.

Overall, there is good evidence that there was a window of time—following the Franklin Dam fight in 1983 and peaking in 1990–1991—during which environmental issues were brought into the policy mainstream. It was an attempt to close the national argument between economic and environmental priorities. Unfortunately it was about to end. Former science minister Barry Jones said in a 1992 World Meteorological Day address:

> Green issues were extremely important in the 1980s and contributed to the Hawke government's electoral success in 1983, 1987, and 1990 ... [but]

> In 1991 with economic recession, the political priorities seemed to change. Jobs, jobs, jobs, became the priority and in some quarters there was a cynical reaction suggesting that environmental issues were luxuries which characterised affluent times ... This is a criminally short-sighted view.

There is much uncertainty 25 years later on whether the Hawke federal Labor government was genuine in its concerns about environmental issues, including the greenhouse effect, or merely catering to an electorate with significant numbers of green voters flexing political muscle. Whether this matters, or is actually a chicken-and-egg debate, the public stance of the prime minister and key government ministers underscores the role of leadership when it comes to controversial public interest issues like climate change.

(Such issues are controversial in the opinion of those sectors that deem themselves to be economic losers when policy responses seem to favour the general public interest. That these sectors fight back by attacking the science behind the policy was also apparent in the decade-long struggle over the hole in the ozone layer that preceded the greenhouse battle. Similar forces surrounded tobacco in that public health policy battle).

Underscoring the leadership role, one senior political journalist wrote in *The Sydney Morning Herald* in September 1988 about federal politics at the time:

> The greatest problem facing Australia today is not its external debt. It is clearly and undoubtedly the environmental threat posed by the Greenhouse Effect. This is not the view of some loony fringe greenie or Australian Democrat; it is the view of Senator Graham Richardson, Labor's right-wing hard man and colleague of the Treasurer, Paul Keating. (Seccombe 1988: 17)

Mike Seccombe wrote that Graham Richardson (then environment minister) understood the big picture is not just the immediate economy, as Paul Keating, his colleague and soon to be prime minister, believed. The big picture might demand considerable change and upheaval in the Australian status quo. He wrote that amongst the steady stream of information crossing the environment minister's desk was the cost of environmental damage including ozone and greenhouse-induced climate change—an annual estimate at that time of $5 billion for the United States alone.

Nevertheless, Seccombe was another observer who in 1988 thought the Hawke ministry as a whole did not comprehend the magnitude of the problem and was not seriously looking for answers—an assessment seconded in recent interviews with John Kerin and Barry Jones, who were in the ministry at the time, and also by Bob Chynoweth, the federal member who acted as liaison between atmospheric scientists and the politicians during this period.

Kerin witnessed the effect of leadership changes. He was minister for primary industries and energy under Hawke in the mid-1980s to 1991, and minister for trade and overseas development under Keating between 1991 and 1993. He co-signed with environment minister Ros Kelly the media release that heralded Australia's emission reduction planning targets in October 1990 (Commonwealth 1990).

In response to a question about early leadership on the issue of climate change, Kerin recalled: 'I grew to have enormous respect for Hawke and felt his awareness of environmental matters was real and deep.' In Kerin's view, Keating, as the next prime minister, relied more heavily than Hawke on the advice of economic rationalist economists in Treasury. Surveying the ministry as a whole Kerin says, 'I didn't think we understood at all the implications of climate change'.

In July 1989 Hawke issued his call to action on climate change in the environmental statement *Our country, our future* (dubbed 'the world's greatest environment statement' by the media). As we've already seen, Australian governments, state and federal, were working towards a national emission reduction strategy.

Led by the prime minister's office, the federal government set up the National Greenhouse Advisory Committee in April 1989 as part of the National Climate Change Program. The committee, chaired by biologist Henry Nix from The Australian National University, was comprised mainly of researchers whose mandate was to fund further scientific enquiry.

A prime ministerial working group under the ecologically sustainable development banner was established at this time with government, community, environmental, union, and business representatives briefed to outline achievable domestic emission reduction strategies, primarily in the energy sector.

Meanwhile, the Australian and New Zealand Environment and Conservation Council (ANZECC) supported state environment ministries and monitored and recorded state-based emission-reduction strategies as they were developed, documenting the extensive revisions to status quo energy consumption as well as native vegetation protection and tree planting being considered.

Who was advising the leaders?

Despite setting up such formal structures, the practice by senior politicians of relying on minders, advisers, and old friends rather than on expert groups or departmental advice was well advanced by the late 1980s, as Nix observed in an interview. This would have an effect on greenhouse policy. Much of the advice may have been on how to stay in office through the next election cycle, but the question of 'who' is within the advising network became an important factor and one often unseen to outsiders.

In Hawke's case, the advice of the chief scientist at the time—Ralph Slatyer, an old acquaintance of the prime minister—was influential, and Slatyer reportedly took a keen interest in climate change. The scientist influence on British Prime Minister Margaret Thatcher was said to be similar. In Australia, strong advice to act on certain environmental issues coming from a highly influential political operator, such as Richardson as environment minister, was also clearly significant.

A contrasting example, leading to a different outcome comes from the lead-up to the Kyoto Protocol international meeting in 1997. One of the most influential voices on climate change policy in the federal government of John Howard at the time was a man described as Howard's 'former flatmate' (Hamilton 2001). Warwick Parer, then minister for resources and energy, was a long-time veteran of the coal industry who was an open greenhouse sceptic and a tireless booster of coal as the cornerstone of Australia's prosperity. He was also alleged to be the minister responsible for abolishing the federal government's alternative energy research corporation in 1998. A year later he resigned from Parliament, charged with conflict of interest due to his coal holdings (Hamilton 2001).

Influence also came from the bureaucracy. Trade officials who crafted negotiation documents and strategies for international discussion were also 'true believers' in the market ideology that came to dominate the 1990s, according to policy scholar and author Clive Hamilton. Market-focused ideology dismantled ESD policy, disbanded the working group briefed to rein in emissions and find efficiencies, and later brought a 'virulently anti-European' perspective to climate policy. From this perspective, Australia's role as a resource quarry was indistinguishable from the national interest (Hamilton 2001, 2007).

Hedging turns into retreat

The October 1990 national interim emission reduction target aimed to stabilise greenhouse gas emissions at 1988 levels by 2000, but an oft-quoted caveat was introduced at the same time. This stated there should be no adverse effects on the Australian economy—upon trade competitiveness in particular—in the absence of similar actions by other countries.

The hedging language signalled that, even at this early date, industry lobbyists, free market economists, and trade bureaucrats were winning with an argument that the fossil fuel economy must stay as it is (being Australia's 'natural advantage') and that Australia should take no action until other countries did. This moment has been flagged as the start of back-pedalling in the commitment to action. Backpedalling strengthened during the 1990s into a full retreat, supported by the often-manufactured confusion and scepticism that marked the later 1990s.

The newspaper record sheds light on the early and conflicting understanding within the Hawke ministry of Australia's reliance on coal exports and domestic fossil fuel intensive energy generation, noting all the themes that came to dominate Australia's position in the 1990s. Ignorance, ideological beliefs, and scepticism in government ranks—affecting policy and communication— built during the Keating and Howard governments: unopposed by the strong environmental leadership on this issue that had been exhibited by Hawke and Richardson. Dramatic consequences for the climate change storyline followed.

The effect of leadership changes interplayed early on with a stalled bureaucracy. The federal Department of Primary Industries and Energy had been asked to rapidly modify national energy consumption and production as a response to the 1990 national emission reduction targets. (While the Commonwealth directs energy policy on vehicle fuel efficiency and appliance labelling, it also can lead by example. It can prod state and territory energy supply and demand balance towards efficiency or renewable energy—including offering incentives and setting up major communication campaigns to frame the issue. In the late 1980s, regulation was also still an option).

A critical 1992–1993 report by the Australian National Audit Office (ANAO) into Primary Industries and Energy's response gave yet another confirmation of the early understanding that greenhouse gas emissions are largely due to human burning of fossil fuels and, in Australia's case, often inefficient use. Australia has a poor record of energy saving, said this report in unequivocal language.

'Market research and technical studies indicate there is a significant untapped potential to save money and resources and stem carbon dioxide emissions. We are amongst the world's largest greenhouse gas emitters on a per capita basis. Our cars are amongst the world's most inefficient in terms of fuel consumption' (ANAO 1993: xi). The audit report also described Australia as lagging behind other countries in industrial plant efficiency, in building construction and in public awareness of the need to save on energy.

It also spelled out the federal government's intentions in 1990 to take rapid response action, particularly to champion efficiencies. In response to this urgent brief, however, the audit office found that most of the federal government's response agenda to the 1990 emission reduction target remained in limbo:

> The Department did not fully respond in the manner expected ... Right up to the announcement in October 1990 it had not fully anticipated the greater priority to be given to the subject ... Staff meant to be available for putting the programs into place were heavily engaged on other tasks such as coal research grants, policy development and advising ... this was the case right until the time of our audit, more than two years after the announcement. (ANAO 1993: xi)

Kerin, who relinquished the portfolio early in 1991, recalls saying to the junior minister for resources:

> There have got to be hundreds of ways of attacking this issue so, for God's sake, do something about local government and design and buildings and power saving ... But I don't think he got anywhere because a lot of this area was in state hands and you know how hard it is to get this Federation to work.

An article from that time in *The Sydney Morning Herald* noted that the Department of Primary Industries and Energy advised Prime Minister Hawke that it would take 30–40 years to make industry more efficient, as old plant had to be replaced. It highlights the emerging framework, which would rule for the next 20 years, that no mandatory changes would be required of large industrial firms, or that pollution costs could be imposed.

> Almost a third of our CO_2 emissions come from just 60 large firms—capital-intensive, using equipment with a long economic life, and export-oriented. The costs of re-equipping would hinder export competitiveness. The alternative of exempting those industries would leave a disproportionate share of the reductions to fall on other sectors. (Seccombe 1990b: 15)

Another Seccombe 1990 report, appearing on page one of the *Herald*, is worth quoting at some length as evidence of the influence of federal Treasury economists and again showing the context of what was known:

> The Federal Treasury is determined to block moves by the Government to make industry cut down on greenhouse effect gases.

> On Monday, when Cabinet meets to consider targets for the reduction of greenhouse gases, the Treasury is set to attempt to delay the matter for up to a year by demanding a new inquiry into the problem.

Government sources believe the suggested inquiry will be a cost-benefit analysis by the Industry Commission of the likely effects of curbs on industry.

Such a move would stymie a submission by the Minister for the Environment, Mrs Kelly, for the immediate imposition of a target reduction of emissions by 20 per cent by 2005.

The Treasury would not confirm that its preferred course was referral to the Industry Commission, but said it believed that no conclusive reports on the value of targets had been done, and more investigation was needed. *A promise of quick action on establishing greenhouse emission reductions was a key plank of the environment policy which played a major part in the Government's election win this year.*[1] [my italics, highlighting the repeat patterns of public knowledge and election promises.]

The Treasury view also ignores the United Nations Intergovernmental Panel on Climate Change (IPCC), which says drastic cuts in emissions of about 60 per cent are needed for the problem to be stabilised. (Seccombe 1990a: 1)

Despite that warning, 24 years later it seems there remains plenty of industry opposition to forcing efficiencies and lowering emissions. A top priority of resource-industry supported think tanks and a likewise supported and like-minded conservative federal government after 2013 has been to drop the previous government's carbon price/tax leading into an emissions trading scheme as well as an attempt to backpedal on renewable energy programs.

In 2013 the issue was again successfully framed in cost terms: as imposing an unfair tax or costs both on industry and on the hip pocket of the mainstream, disregarding that the risk-management objective for everyone is emission reduction.

In contrast, going back 20 years, strong intentions to respond to climate change with a suite of measures were recorded well into 1992, culminating with Australia's participation at the UN 'Conference on Environment and Development' (also known as the Rio Earth Summit) and the simultaneous establishment of the UN Framework Convention on Climate Change (UNFCCC). The Australian delegation, including Kerin, was headed by then environment minister Kelly, who was already fighting the departmental and bureaucratic turf battles at home.

---

1   In 1990 and 2007 promises of action on climate change were credited with helping swing a federal election—only for the promised action to eventually evaporate. In both cases the electorate was credited with being aware and eager for action.

State and territory action plans were well underway by 1991. The titles are unambiguous. For example, in 1990 the ACT Government released a document *Developing an ACT strategy to respond to the greenhouse effect*. The Northern Territory was working on a plan, and every state was refining one. The answers the states were devising involved regulation and incentives for efficient energy use in the residential, commercial, and industrial sectors, as well as boosting areas like public transport. States promised to gear up for renewable energy programs. Victoria and Western Australia instituted 'major' energy demand management programs to lower energy consumption (ANZECC 1991).

Most of these programs were destined to die on the altar of deregulation, competition policy, and free market ideology in favour of purely 'supply' options (meaning opting for more consumption), in succeeding years. At the same time there was a major narrative shift from risk management to cost management.

Reframing to favour 'business as usual', themes that still play

The late 1980s science information didn't 'dissipate'—it was blasted away.

Engineer and energy consultant Deni Greene

In 1990, energy analyst Deni Greene was commissioned to prepare a number of analyses for the Hawke and then the Keating government's Department of Environment and the ESD working groups. The brief was to identify opportunities to curb energy demand, estimate related dollar savings, and explore job-creating new industries for energy production (Greene 1990a, 1990b). Her reports offered evidence that energy efficiency and conservation alone could meet Australia's interim emission reduction target and that it was a 'win-win' way to save money *and* the environment.

Some observers at the time suspected that the resource and energy industry associations complained about these publications to their contacts in government. Greene told me that the environment department subsequently drew back on commissioning her work. This episode may be early evidence of the unusual nature of the late 1980's mainstreaming of environmental messages, and that a more traditional public narrative was about to return. This would be framed by politicians, bureaucrats and the media as the need to defend existing industries and markets while downplaying or ignoring the 'win-win' of emission reduction with new energy industries and related job creation.

The traditional narrative pitted environmental action *against* the economy and jobs, an 'either-or' argument. The classic example was the long-running native forest dispute. The realisation that preventing greenhouse gas pollution might incur costs was used strategically in the 1990s to help frighten and manipulate the public into uncertainty about the science and about the need to respond urgently.

A closer look at how this happened shows that by 1992 the associations of extractive industries began to stir, led by coal and oil and multinational processors particularly aluminium (lured to be reliant on 'cheap' coal-fired electricity), which stood to lose from any change in existing energy production. Backed by free market economists, they were seeing the implications for their 'business as usual' operations and their fightback began (Pearse 2007), starting with challenges to the scientific risk assessments.

Through industry documents and political allies, the public discussion was turned to a focus on costs and jobs, and also free market and 'choice' ideology. The evidence indicates that this counter argument gathered steam in response to Australia's interim emission reduction target and the recommendations for global action of the 1992 UNFCCC.

How it was framed for the public can be seen from documents produced by CRA on two occasions. (Unpack the acronym CRA and a familiar entity emerges— Riotinto, incorporating Anglo-Australian mining giant Conzinc Riotinto of Australia, CRA). A 1989 CRA report on Australia's proposal to lower emissions by 20 per cent established some now familiar themes: it was alleged that there would be little global effect and it would damage Australia's economy; holding down demand (through efficiency measures) would be costly and lead to unwelcome lifestyle changes; battling climate change is just fear of change; and, warming trends may benefit some agriculture (Dixon & McLennan 1989).

A 1992 CRA report in the *Mining Review* warned of severe economic consequences if Australia implemented the UNFCCC commitments to lower emissions. The article rests on the 'us and them' scenario (Carruthers 1992). A frame that would become familiar was anti-UN and anti-European rhetoric claiming foreign forces were trying to tell Australia what do, and were damaging Australia's national interest. The United Nations and Europe were seen as the main drivers for binding emission targets.

This report is an early example of the soon commonplace strategy of quoting Australian Bureau of Agricultural and Resource Economics (ABARE) statistics that were then amplified by the media and seldom questioned. Thus in this article, ABARE is quoted as putting a likely carbon tax at $160–200 per tonne, which would damage export industries and occasion massive job losses

(Carruthers 1992). (In 2014, the actual carbon tax was around $24 per tonne.) ABARE's director Brian Fisher emerged during the 1990s as a reliable sceptic voice in regard to economics and the cost of response activities.

In fact, the public relations work of CRA in this 1992 report established the contrarian themes that came to dominate the decade. It featured:

- scepticism—it is not the fault of humans
- loaded images—'who is behind the greenhouse panic?'
- market ideology—those who push the greenhouse science do not believe, as do the rest of us, that everyone should have options and be able to make choices
- 'us and them'—much of the 20th century has been a struggle over two fundamental positions: freedom and choice on one side, Marxism, fascism, and religious fundamentalism on the other side—'such people have now discovered environmentalism'
- suspicion of scientists' motives—they are just after grant funding
- contention that the media is being manipulated by 'the totalitarians and less-than-scrupulous scientists'
- belief that acting to conserve biodiversity is against development interests
- the argument that acting on climate change would have little global effect, but large costs, for Australia.

Shift from risk to cost management and delay response

Politicians stopped talking about the need for urgent risk management of climate change and started talking about cost, uncertainty and that there was no need for immediate action. This framing started emerging alongside the economic downturn that came to dominate public discussion by 1991. The Australian Government's response, by then under Prime Minister Keating, is described in an October 1991 business report in *The Sydney Morning Herald*. Noting the Australian penchant for delay via reports and commissions, journalist Paul Cleary wrote that the most recent attempt to delay serious action was by referral of the problem to the Industry Commission. 'The former Treasurer, Paul Keating … was one of the prime movers in deferring a final decision on the now-famous Toronto target, a 20 per cent cut in emissions from 1988 levels' (Cleary 1991).[2]

---

2   Cleary, however, was impressed that the Industry Commission had developed something called the World Economic Degradation General Equilibrium (WEDGE) model specifically to crunch the costs of responding to climate change with 270,698 variables and 245,831 equations (Cleary 1991) —an interesting example of how economic number crunching can be accepted uncritically in media analysis.

Analyst Deni Greene saw Australian industry by the early 1990s as a cohesive voice fighting the science and potential response messages. She said that a lot of companies could have benefited economically from efficiency measures, but did not speak out. She concluded that this silence was based on a business 'kinship' response, and the possibility that businesses 'bought' the cost argument uncritically. Industries that stood to lose from climate change, like tourism and insurance, were equally silent or acquiescent to the resource sector.

Like other contemporary observers, Greene also saw the leadership of Paul Keating as weak or disinterested as far as energy policy and climate change were concerned. Like many of those interviewed about this period, she saw him as focused on economic matters, starting with response to the 1991 recession (the famous recession 'we had to have'), and also with economic rationalist reforms. Keating himself has told other interviewers that he has acted as a defender of the natural environment and threatened ecosystems and landscapes in Australia, some of which indeed received federal protection during his tenure as treasurer and then prime minister.

Us versus them becomes new normal

The long-running native forest dispute—involving government-subsidised rural industries extracting natural resources supported by both major political parties, versus some members of the public—helped forge a blanket negative stance from rural and extractive industries (and the politicians who represented them) against environmentalists and environmental groups (Mercer 1991; Pearse 2007; Ajani 2007). Environmental campaigners were labelled extremists who were against the necessary balance between the economy and the environment. They were therefore cast as special interests that did not have jobs and the national interest at heart.

As the 1990s proceeded and industry groups joined forces to combat climate change action, this framing of environmental concerns as being against jobs and the national interest became the new dominant narrative, or more accurately, an old narrative renewed. This world view became the new normal as politicians and the media decided the climate change story was primarily political and economic.

This view gained more currency as green organisations took over from scientists in media quotes pointing out the risks of climate change and opposing Australian Government policy on the subject by the mid-1990s. It was then a short step to painting them as a 'special interest'. Evidence for this change towards adversarial politics can be seen in media reports from the time and is particularly evident in business reporting, as in *The Australian Financial Review*, which became more industry partisan compared to just five years earlier.

An analysis of 30 articles in the *Financial Review* between 1995 and 1996 showed how the reporting shifted towards support for the concerns of existing industries. The following example that introduces an article is one of many from the later 1990s showing how Australia had changed its story and where it was headed: 'Australia's push for international greenhouse policy to focus on economic issues, rather than narrow technological or environmental targets, is likely to receive a warm hearing at a major conference on climate change in Geneva' (Callick 1996b).

At that time, the *Financial Review*'s later role as an uncritical advocate for the resource industry was still only partial. In the mid-1990s reporter Michael Stutchbury (more recently the editor of this publication) continued to dissect the economic debate in a revealing manner. For example, he reported that Australia's argument on the international stage for exceptional treatment when it came to emission reduction was because it wants to continue trading unabated in fossil fuels. On the other side of that debate were emerging economies arguing that those who had already polluted the globe should bear responsibility and clean up first (Stutchbury 1995).

Carbon tax already a scare campaign in 1995

Greenhouse topics dominating the business press in 1995–1996 were the spectre of carbon taxes and industry's opposition to them, and Australia's opposition to global mandatory emission reduction targets. This was made clear at the March 1995 Berlin conference of the parties to the UNFCCC, which preceded the Kyoto meeting two years later. Australia signed the UNFCCC in 1992.

A 1996 *Financial Review* article, after the federal government had changed to the conservative parties under Prime Minister John Howard, illustrates the dominant political climate and narrative and also the revolving door of familiar players: thus is worth looking at in more detail. Under the headline 'Business lines up to fight controls', the report said:

> Business has warned the Howard Government to reject international proposals for a raft of new taxes to combat greenhouse gas emissions.

The president of the Business Council of Australia, Mr Ian Salmon, has this week written to the Prime Minister, Mr John Howard, and other ministers calling for a tough stand against accepting legally binding, uniform targets for greenhouse emissions.

The executive director of the Minerals Council of Australia, Mr David Buckingham, wrote to the Minister for Resources, Senator Warwick Parer, registering 'strong concern' that a briefing last Thursday was 'the first attempt by the Government to engage industry with the detail of the proposed Australian approach to this critically important treaty negotiation'.

Mr Howard last night assured the Minerals Council of Australia that industry and non-government organisations would be given proper involvement in international lawmaking on the greenhouse issues … 'It will insist that Australia's economic and trade interests are safeguarded and its specific national circumstances are taken into account in implementing the convention.'

[Mr Salmon of the Business Council] said research by the Australian Bureau of Agricultural and Resource Economics indicated that a harsh approach to industrialised countries like Australia 'would have little environmental impact given the unconstrained growth in developing country emissions, especially those of China, India and Indonesia in our region.'

Mr Buckingham said: 'Any outcome from the climate change treaty process that results in a 'ramping up' of existing targets and timetables would be fundamentally contrary to Australia's economic and trade interests.' (Callick 1996a)

In framing terms, scare rhetoric—when speaking of a price on carbon pollution— and wielding that negative word 'tax', was well entrenched by the mid-1990s and continued into the 2000s: reaching a crescendo of accusation as a political weapon wielded by the conservative Opposition against the Labor government under Julia Gillard after 2010. Other themes that were well developed, judging by those mid-1990s articles in *The Australian Financial Review*, told the public: if Australia institutes targets and a timetable to combat carbon pollution, it would scare investment away; Australia would lose its competitive advantage; Outsiders, Europe and (at that time) the United States were doing this type of damage to the country from motives of trade competitiveness and internal politics (in the case of the United States); jobs were at stake.

As we see from the preceding excerpt, the Business Council of Australia and the Minerals Council were prominent in setting these narrative themes.

Spokespeople were part of the Australian Industry Greenhouse Network (AIGN): a revolving door of former bureaucrats and industry lobbyists who influenced John Howard's thinking on climate change action (Pearse 2007).[3]

During the same period *The Sydney Morning Herald* remained more diverse in emphasis. But several environmental journalists who worked on the *Herald's* parent company Fairfax's metropolitan papers after the mid-1990s said that science and environment stories always ranked lower than economics or politics if it came to an editorial choice. This was also true in the United States (Gelbspan 2004). Environment had reverted to low status compared with the traditional arenas of competitive politics, economics and sports.

Newspaper reports from this time mostly quoted green groups as spokespeople for climate change action in the political reports. After 1996 they were usually quoted in opposition to the government line at international gatherings, where it was argued that Australia was exceptional and should not need to extensively cut emissions. This painted green groups as being opposed to jobs and the national interest: another story about 'us and them'—that is, mainstream versus selfish special interests.

It has been argued that environmental groups have been equally intent on maintaining an adversarial role apart from the mainstream, and on advocating narrow legislative and technological solutions (rather than, for instance, focusing on the opportunities for alternative job creation), thereby ensuring a narrow frame of influence within society (Shellenberger & Nordhaus 2005).

Media amplifies the new narrative, opinion gains ground

During the second half of the 1990s, government-level communication about climate change became more technical and 'boring' to a lay audience, moving away from the earlier mainstream discussion of risks to every household posed by greenhouse gas emissions and climate change. The change from the direct and accessible to the narrow and technical can be traced through the content of the federal government's *Climate Change* newsletter, which was published between 1992 and 2000 by the Department of Primary Industries and the Bureau of Rural Resources.

---

3 Guy Pearse in his 2007 book *High and Dry* documents how in Australia there has been revolving door amongst former senior bureaucrats who then head industry lobby organisations like the Minerals Council of Australia and other arms of the fossil fuel lobby. He quotes one source who says of the former industry, energy or Prime Minister and Cabinet staffers: 'We all write the same way, we all think the same way, we all worked for the same set of ministers' (Pearse 2007: 230). In the 2000s many still revolve in executive positions including in higher education.

The national broadcaster, the Australian Broadcasting Corporation (ABC)—which in the early 1990s had led the way with comprehensive science coverage of greenhouse gas emission risks and response strategies—became increasingly dominated by a conservative board of directors appointed by the Howard government (Dempster 2005). The ABC then largely restricted itself to amplifying the political story spun by the federal government (or the opposition), reporting issues framed through the echo chamber of the parliamentary press gallery.

Australia's only national newspaper, News Limited's *The Australian*, took a largely sceptical stance during the 1990s and has continued that stance to the present day. News Limited tabloids in every major city took a similar line and have only become shriller in the 2000s. Media analysts have linked this to the strong market fundamentalist editorial stance held by management, where action on climate change was seen as a threat and an unwelcome cost to doing business (McKnight 2005a).

Climate change coverage at the editorially more neutral *Sydney Morning Herald* also changed by the second half of the 1990s. A sceptical 'balance' was introduced when discussing the science, often deploying armchair sceptics who countered news reports of a scientific development. This trend also reflected a 10-fold increase from 1988–1989 to 2001 in opinion pieces by non-staff writers on the subject, along with a shift away from always using scientists and experts as the primary source of information.

In the concentrated Australian media market (with one major player, News Limited, running a sceptical line and the business press becoming partisan) these changes had significant influence on the dominant narrative. It increasingly made a discussion about climate change appear to revolve around opinion and belief rather than evidence.

Connections to free market think tanks also played an important role, with think tank members often writing opinion pieces for News Limited papers (Manne 2011; McKnight 2005a). The think tanks also had close links with the minerals industry, not least through funding.

A better understanding of the influence of the resurgent beliefs and ideologies promoted by these sectors through political leaders and media is fundamental to understanding what happened to climate change knowledge in Australia in the 1990s and since. I look at these beliefs and ideologies in greater detail in the following chapters as well as the media influence. The most influential framing of the climate change narrative was not done by either politicians or media stories alone, but occurred when there was agreement between political and media narratives and across different media platforms as media outlets echoed each other.

By 1996–1997 political and economic reporters and editors in the parliamentary press gallery were dutifully scribing the story established by the business and political elite: dealing with climate change was all about a political struggle to get the world to accept that Australia was exceptional—because it traded heavily (with often inefficient technology) and offered coal-fired electricity to energy-intensive multinational companies and to Australian consumers. There was of course also the conflicting matter of being the world's largest coal exporter. Any change from this status quo was not acceptable.

Those who did not agree—the environmental groups, other countries including vulnerable Pacific Islands, Europeans, or the United Nations itself—were framed as the opposition to Australia's growing prosperity enjoyed by the mainstream, particularly Australia's aspirational voters and 'battlers'. In 1997, the prime minister put it this way: 'We are not prepared to see Australian jobs sacrificed and efficient Australian industries, particularly the resources sector, robbed of their hard-earned competitive advantage' (Howard 1997). The speech in fact canvassed response activities that the government proposed to take to the upcoming Kyoto negotiations, but the public heard metaphorical messages of theft and 'it's unfair'.

Bureaucracy power plays and trade wins

Institutional factors also helped the reframed story take over the public discussion. Since leadership had a significant influence on what happened in the 1990s, it follows that leadership style also influenced the way the response was tackled. While Hawke's style as prime minister was consensual, bringing all sectors to the table, RMIT alternative energy expert Alan Pears witnessed the policy transitions from the early 1990s on, and said in an interview about the next Labor prime minister:

> Keating's style was bureaucratic. Climate policy became fractured between 37 committees of bureaucrats … by 1994 a number of the threads started to coalesce that killed off ecologically sustainable development work while industry leaders and most of government thought 'supply side' i.e. more energy development [rather than conservation] equals growth and development.

The Sydney Morning Herald reported on how this was shaping up in the early Keating years:

> So far the opponents of the [emission reduction] targets have employed the favourite trick of the bureaucrat—delay—to bog the whole process down in a myriad of inter-departmental committees, studies

and consultancies. The three ministers with central responsibility to implement the changes to reach the target—Kerin (now Crean), Beazley and Button—were asked by Cabinet last October to report back by the end of last year on 'recommended implementation measures'. Nine months later and there is little sign of them rushing back to Cabinet. (Burton 1991: 32)

The federal energy bureaucracy's internal priorities were elsewhere as we learned from the 1993 auditor general report (ANAO 1993). Interdepartmental battles did not help either. Like other insider participants interviewed about the 1990s, Kerin observed that the native forest debate poisoned relations between the Department of Environment and the stronger Department of Primary Industries and Energy that he led until 1991, and recalled the frustration he felt working with green groups, some of whom he still accuses of having lied about forestry issues.

Trade and Treasury were also involved in these battles over the forests. Along with the influence of market ideology and industry lobbying campaigns, the resulting antipathy from finance and resource industry bureaucrats to environmental action significantly bogged down the early climate change action plans by the mid-1990s under the Keating federal government.

Sue Salmon, from a green non-government organisation background, was an adviser to environment minister John Faulkner (1994–1995). She also experienced an internal fight between the primary industries and mineral extraction portfolios and the environment portfolio, and agreed that forests remained a major focus for the environment movement during those years. Climate change was considered a difficult issue to communicate in an ongoing fashion compared to forests.

By then the head of the federal Department of Foreign Affairs and Trade oversaw climate change action proposals. Salmon said in an interview it became all about 'the traditional conservative view: we have hundreds of years of coal to trade … It was very "us and them" and there was a perceived loss of power and face and control by the industry groups to accept the environmental perspective'. She also recalled that the IPCC was not viewed as an important avenue for information within government.

Other contemporary observers, such as Henry Nix, similarly perceived the environment portfolio to be weak. During the time he was chairing the Greenhouse Advisory Committee, Nix said the environment portfolio was generally at the losing end of this argument. 'Even on the best days economic arguments always prevailed. It was possible to modify but not change much about it,' he told me. This was consistent with the low status afforded environment in the mass media.

Phillip Toyne was at the coalface and agreed that climate change lost out in the policy debate. He had been drafted from the leadership of the Australian Conservation Foundation to deputy secretary of the federal environment department in 1994. He said in an interview that looking at some ice core research in Antarctica 'woke him up', but that generally neither environmental groups nor government had climate change at the top of the agenda in the 1990s. Environmental and scientific submissions might prevail to a certain degree, but only for a short while. In his view there was 'a major disconnect between what scientists knew and their ability to influence policy [and this was due to] internal CSIRO traditions that did not promote a lot of communication with policymakers'.

During Toyne's tenure (continued for two years under the Howard government) he found the government's scientific advisers to be 'invisible', certainly not wielding any direct influence. Toyne agreed that during the Hawke/Richardson leadership on environment matters in federal government, the environment generally and climate change in particular had been treated as 'mainstream'; that is, of concern to all citizens. But by the mid-1990s, with the industry lobbyists in full swing, the federal government was treating climate change knowledge as a 'special interest' and a not very welcome one.

In these ways, political leadership style—along with leadership intent—and bureaucratic beliefs and values played key roles in framing how to think about climate change at the national level from the Hawke through the Howard governments.

From the mid-1990s, industry lobbyists increasingly flexed their muscle under the umbrella group the Australian Industry Greenhouse Network. Salmon said they were very effective talking about jobs and income creation, and Australia's 'national interest', while the environment ministry still focused on degrees of certainty.

Underlying this winning economic story were values and beliefs that elite politicians, bureaucrats, and industry leaders tended to have in common, particularly about Australia's role as a quarry to Asia and the transcendence of macro-economics in all policy formulation. Kerin recalled:

> Keating and (later Howard government Treasurer) Peter Costello were suckers for dogma on macro-economics … The herd instinct (became) 'the market, the market, the market'. In my terms they never examined enough market structure, market power, market failure

> One of the things that has always worried me about economics and science, (and I set up the bureau of rural sciences in my department, because all they were concerned about was trade and economics) is that

economists are always absolutely sure they are right. Even when they're subsequently proved wrong they just forget about that. Scientists are never absolutely sure they're right because they always know there's more discovery and we learn more and more.

A similar observation was made about the trajectory of the NSW Labor government under Bob Carr when Salmon worked there as an advisor. She noticed a similar weakening split within government ranks with Carr apparently understanding the science, but the Treasurer Michael Egan blocking and presenting arguments as a sceptic and economic rationalist.

From key reforms to policy sideline

In taking a closer look at the fate of the now historic ecologically sustainable development (ESD) process, it's clear how the same scientific and economic data can lead to widely divergent policy recommendations and public storylines. It also shows what can happen to good ideas or programs that lack the leadership to ensure they survive. ESD's eventual sidelining indicates the difficulties of partial change, let alone radical revision of modern industrialisation and the institutions it relies on.

ESD was then a global concept attempting to add environmental and social justice benchmarks to planning and development. It was put into practice by the Hawke Labor government's 1989 decision to involve governments, industry, environmental, and community representatives in working groups. They were to consider nine sectors of the economy, including resource and energy use and assess community wellbeing, intergenerational equity, global impact, protection of biodiversity, and ecological processes along with economic development (Harris 1997). The ESD experiment gave equal standing to the natural environment and its spokespeople.

A combined ESD working group was to recommend how to lower carbon emissions from the energy sector. In 1990–1991 its priority recommendation was to focus on efficient energy use, thereby also lowering costs: so-called demand management (Bulkeley 2000b; Diesendorf 2000). This could be done by advocating or regulating smarter ways to operate in residential, commercial, and transport sectors: substituting gas for electricity; using insulation; using efficient motors; and promoting better construction and planning as well as changes in the agricultural sector to create greenhouse sinks (more recently called bio-sequestration or carbon farming).

According to former science minister Barry Jones, '[This] approach begins with the assumption that something can be done, that the argument "this is the way we have always done things around here, and it can't change" is unnecessarily pessimistic' (Jones 1992: 7). Compared to political debate today, the wide range of solutions canvassed were remarkable also for not being an either or proposition (a carbon price or direct action).

The 500 recommendations that emerged from this ESD assessment did not challenge the basic assumptions of modern industrialisation and growth. The degree of consensus, however, including from business and non-government groups, was surprising reported British geographer Harriet Bulkeley, who looked closely at this period in Australia. In the event, ESD still proved too radical a process to last. Australia's elite decision-making tradition emerged to drive the outcomes in the transition between Hawke and Keating. This was achieved by the federal government developing the initial discussion paper and terms of reference; by the numerical preponderance of bureaucrats in the working groups, and through the government's selection of the stakeholders from industry, environment, and community groups.

The important energy sector report with implications for greenhouse gas emissions was taken 'in house' in 1991 where interdepartmental and intergovernmental committees whittled down the recommendation list to form the basis of the National Greenhouse Response Strategy (NGRS) and the National Strategy for Ecologically Sustainable Development (NSESD). 'The resulting draft NGRS bore few similarities with the conclusions of the working group, representing instead a "lowest common denominator" approach as to what governments and bureaucrats were prepared to accept' (Bulkeley 2000b: 42).

The cross-sector ESD experiment ended in discord and disarray. The national Institution of Engineers, not known for its radicalism, issued a press release in August 1992 condemning the process, saying: 'According to the Institution, bureaucratic arrogance in the National Greenhouse Steering Committee (NGSC) has produced a National Greenhouse Response Strategy (NGRS) which encourages procrastination on all actions—even on those measures which are well-proven as being cost effective' (Dack 1992).

The response themes that would come to characterise the 1990s and 2000s were evident in the national strategy that the Institution of Engineers criticised: delay, more research, voluntary action and rejection of demand management and mandatory efficiency measures—leading to the eventual burial of the national 20 per cent planning target for emission reduction.

Triumph of the economist world view and 'can't do' climate response

Then treasurer, soon to be prime minister, Paul Keating had asked for those ESD recommendations, along with a similar request to the Industry Commission for comparison. Both organisations were to look at the costs, benefits, and opportunities of the government's draft emission reduction target. The eventual triumph of the Industry Commission assumptions and world view would direct a decade of responses to the scientific information about climate change and provide plenty of ammunition for the new public narrative. While the ESD recommendations were whittled down and homogenised, the simultaneous Industry Commission analysis of costs and benefits came to dominate the responses suggested in the NGRS (Bulkeley 2000b).

In his 1992 World Meterological Day address Barry Jones candidly compared and contrasted the 1991 ESD and Industry Commission reports on costs and benefits of greenhouse action. He showed that, in many ways, the two analyses reflected a 'can do' versus a 'can't do' view of effective response to climate change. The 'can't do' framing helps explain how Australia embarked on two decades of losing valuable time to act (Jones 1992).

As a premier research organisation for the Australian Government, the Industry Commission initiated its inquiry in January 1991 with public hearings and submissions from 'interested parties', which is a common Australian practice. The commission reported in November of that year (Industry Commission 1991). Given the 'business as usual' and growth assumptions it entered into its modelling, the commission found that there would be a heavy cost to Australian industry if emissions were corralled at 1988 levels by 2000 and reducing 20 per cent from there by 2005. This would require some changes from the status quo of Australia growing as a 'raw materials economy' and would affect existing industries like coal and oil.

In his address, Jones called the commission's approach 'rigid' and wedded to recurrent ideas of the nation's 'comparative advantage' (a neo-liberal market economic idea) as a quarry to developing countries, and a global base for energy-hungry industries like aluminium. It was noted in contemporary media reports that, despite its good intentions on climate change mitigation, the Hawke government was keen for this to happen, rationalising that the coal was 'low sulphur'.

A 1990 *Sydney Morning Herald* article pointed out that this view was not unanimous in the federal Cabinet: 'That argument [to entice industry to Australia with its coal-based electricity] has potential merit, except that, as (Environment

Minister) Mrs Kelly points out, Australia has the least energy-efficient industrial sector of any OECD nation—that is, it must burn more fossil fuel to achieve a given industrial output' (Seccombe 1990b: 15).

An inefficient industrial sector in an export environment helps explain the cost assumptions for the Industry Commission, and its findings of cost and hardship supporting the 'can't do' theme. Australia during these years also made fundamental choices not to have a diversified economy, thereby limiting its response options (Pearse 2009).

The 'can't do' story also got an unexpected assist following the plain-English 1990 IPCC report. After this communication high point, many climate scientists started to take a more cautious and conventional public stance that stressed uncertainties, returning to more familiar communication territory in response to public attacks at the IPCC and domestically. Changes in scientists' own language became another important influence on the eventual climate of uncertainty besetting Australians that continues to the present.

The Industry Commission issues paper foreshadowed several other theme shifts as the 1990s progressed—eventually away from the idea that there is a global ethical responsibility and toward the frame that the global atmosphere is a commons, and no one nation can have significant impact unilaterally 'if in doing so this significantly damages their economies or international trade competitiveness' (Industry Commission 1991: 5).

Oft-repeated thereafter, was the threat that if Australia acts against major industry emitters, they will simply move offshore (Industry Commission 1991: 12). With these frames, still very current today, the Australian narrative came to illustrate Garrett Hardin's classic thesis of the 'tragedy of the commons'—where no group is willing to unilaterally look after the common interest, thinking that others will not.

Fixed idea: markets are perfectly efficient

The Industry Commission developed economic modelling that set a benchmark for other government advisers, particularly ABARE and its industry clients. The modelling of costs to the economy provided the political ammunition for not ratifying the Kyoto Protocol to set global emission reduction targets after 1997. Trumping the ESD recommendations to mandate increased efficiency and switch to alternative energy or fuel mixes, was a world view about the nature of markets: that they are inherently efficient.

With that idea, regulating economies for social outcomes, like mandating fuel efficiency, or even offering incentives for change was undesirable. This ideology is part of economic rationalism as practised in Australia, a major influence on response since the 1990s, which is examined in more detail in chapter 6.

In an analysis of what he calls 'idealist economics'—that is, theory and policy divorced from on-ground evidence—political economist Evan Jones cited the work of the Industry Commission as a prime example of economic modelling and analysis using 'a preconceived conceptual framework' (Jones 2002). ABARE's analyses in the 1990s looked similarly 'idealist'. Political scientist Guy Pearse interviewed insiders and describes how assumptions favouring status quo industries and the likely cost of any change were seeded into the ABARE's economic modelling of the mid- and late 1990s. Why this was done has been variously ascribed to strongly help beliefs or consultancy payments from industry, or likely both. 'For ABARE, Australia's big polluters are in fact clients. Many of them have paid vast sums for ABARE's greenhouse policy research. The terms of these deals are commercial in confidence, not even revealed in parliament' (Pearse 2007: 219).

Parliamentary records, however, do reveal the biggest names amongst Australian fossil fuel companies as clients who funded the ABARE model that underpinned Prime Minister Howard's climate change response after 1996. These include the Australian Coal Association, Australian Aluminium Council, the Business Council of Australia, BHP, Rio Tinto, Exxon Mobil and other oil companies. Their involvement generated enough controversy to merit an auditor-general's investigation in the late 1990s, which showed no environmental lobbyists were on hand to wield comparable influence (Pearse 2007).

Australia as a good global citizen

Despite this economic frontal attack brandishing cost assessments, Australian politicians were still keen to be seen as good global citizens through the early Keating years. In June 1992, Australia was a signatory to the UN Framework Convention on Climate Change (UNFCCC) unveiled at the Rio Earth Summit. The UNFCCC came into force in 1994 and Australia became the eighth nation to ratify the convention, signalling its serious intent. The UNFCCC called for emission reduction of greenhouse gases to 1990 levels by 2000, and targets would be set in 1997 at the Kyoto meeting of the parties to the convention (the Kyoto Protocol). Signatories were also supposed to design effective response strategies—which Australian governments had been doing since the late 1980s.

Project Victoria, blueprint for undoing the best state emission reduction plan

All Australian states reported in October 1991 on what had been achieved since their mission statements in 1988 (ANZECC 1991). Victoria is the premier example of how far climate change response programs had progressed and what happened thereafter.

At the beginning of the 1990s, Victoria was leading with a comprehensive suite of intended actions including: mandatory insulation in new housing; permanent controls on tree-clearing; incorporating the costs of environmental damage by 'providing a 10 per cent cost advantage to energy conservation and renewable energy resource options'; and (soon to be disbanded by deregulation and privatisation) major demand-management programs through the State Electricity Commission of Victoria (SECV) and the Gas and Fuel Corporation of Victoria. All this would require 'a policy of energy conservation rather than increased sales' (ANZECC 1991: 38).

The SECV figured that its programs—targeted at residential, commercial, and industrial consumers—could save 14 million tonnes of greenhouse gas emissions annually while reducing demand, thereby saving consumers money, up to the year 2005. (Meanwhile, nationally the Industry Commission modelling, and later ABARE's, would disregard the possibilities of demand management in favour of assuming incremental growth in demand and costs).

The Renewable Energy Authority of Victoria was taking even bolder steps to finally crack the builder and subdivision mentality, with programs including guidelines for solar-efficient subdivisions to be incorporated in the building code; house energy-efficiency ratings; labelling for solar hot water heaters, wind farms, and methane recovery at landfills (ANZECC 1991).

All this was to come to a dramatic halt with Project Victoria, which brought a Liberal government led by Jeff Kennett to power in 1992 with the help of a blueprint for a deregulated state and a market economic approach to administration drawn up by the neo-liberal think tanks the Institute of Public Affairs (IPA) and the Tasman Institute (Cahill & Beder 2005). Kennett would lead these 'reforms' until 1999. Ideology was shifting the goalposts at both the state and federal levels. (Reform is a word that signals not just change but improvement. I use the quotation marks to emphasise how language is deployed and what we are likely to hear.)

Deregulation of state energy utilities following 1992 (aided by the Keating government's enthusiasm for national competition policy) helped undo both the intent and the capability for reducing Australia's emissions through energy-demand management strategies. Victoria was not alone in planning energy sector efficiency management. But promoting commercial competition in the energy sector favoured increasing revenue through increasing demand, the opposite of demand management. The response from the other states varied. In some cases, some conservation programs did remain for the longer term. For example, New South Wales created its Sustainable Energy Development Authority (SEDA) that promoted green power electricity and energy efficiency into the 2000s (Diesendorf 2007).

Weak national response a bow to commercial interests

The 1992 national greenhouse response strategy (NGRS) has been called weak and ineffective by more than just the Institution of Engineers:

> The failure of the NGRS derives from a failure of governments to show leadership, to reconcile conflicting policy objectives and to distinguish the public interest from narrow commercial interests. This has been compounded by a lack of knowledge of the energy market in parts of the bureaucracy, and a lack of informed public debate and scrutiny. (Wilkenfeld, Hamilton & Saddler 1995: 1)

Voluntary and ad hoc activity characterised the NGRS. The strategy established a framework that response should be left to individual action and be 'no regrets' (i.e., no entity should bear costs) which became firmly established in the public discussion and assumptions of the possible.

In that spirit, the showpiece of federal government action by 1994 became the Greenhouse Challenge Program, which was targeted at voluntary industry efficiency measures, and administered jointly by the departments of environment, energy and industry. It reflected the federal industry department's decision not to compromise growth and development on behalf of greenhouse gas abatement goals, regardless of international commitments (Bulkeley 2000b: 47). The Greenhouse Challenge Program has been described as a model of how to tackle a pressing national problem in an ad hoc, voluntary fashion—rather like asking citizens to voluntarily tax themselves for the public good.

It reflected the ideas behind most climate change response activities by the federal government as the 1990s unfolded and beyond: that 'the market' knows best on all things. A carbon tax, considered an effective tool in a market economy, was nevertheless rejected as early as the mid-1990s as too costly to industry (Hamilton 2001).

By 1994–1995, it became clearer that Australia would be overshooting the UNFCCC aim of reducing greenhouse gas emissions to 1990 levels by the year 2000 (let alone the earlier domestic interim target of 20 per cent below 1988 levels), and therefore Australia would not be meeting its international commitments or indeed implement much of the NGRS (Bulkeley 2000b: 47; Hamilton 2001).

By mid-1995, the federal government had formally aligned itself with the so-called JUSCANZ countries—standing for Japan, the United States, Canada, Australia and New Zealand. At least one major environmental organisation blamed the United States, Canada, and Australia in particular for obstructing a whole raft of environmental measures that were agreed to at the 1992 Rio Earth Summit, including moving ahead on reducing greenhouse gas emissions (Greenpeace 2002).

While Australian scientists remained involved at the IPCC level, and the scientific message remained similar, the 1995 IPCC report would be a cautious shadow of the 1990 report in communication terms. Reasons offered by various observers have been: behind-the-scenes politicking by oil and energy producers and their alliance with sceptic scientists and the need to build consensus amongst government officials from many countries for the policymakers' summary. IPCC summaries for politicians reflect expert analysis filtered through a consensus process with policy officials; members of scientific panels have noted that this led to a lowest-common-denominator approach that under-reports the risks.

Australian National Party politician John Stone approvingly noted in an opinion piece in *The Australian Financial Review* that the 1995 IPCC report was only '40 per cent as apocalyptic' as its 1990 counterpart. In doing so, he said he echoed the sentiment of US sceptic scientist Patrick Michaels. At this rate, Stone hoped the whole lot of 'poppycock' would disappear by the end of the decade (Stone 1996: 25).

The reframe from 1996 on

The dominant narrative was well on its way to changing from a 'science and risk to society' story to a political and economic story about costs and 'national interest' as the conservative Howard Coalition government took office in March 1996. Newspaper articles provide evidence of the strong economic focus that continued through 1996–2001 (and beyond). The following *Financial Review* article makes manifest the government's identification with resource industry interests, and the language is typical of the later 1990s in this publication. Headlined: 'Coalition backs industry on climate change' the 1996 story reports that:

Australian industry has applauded the Federal Cabinet's decision yesterday to oppose a targets and timetables approach to international climate change negotiations, made on the eve of World Environment Day today. The Howard Government's position effectively reaffirms that taken by the Keating government and its minister for the Environment, Senator John Faulkner. The Minister for Foreign Affairs, Mr Alexander Downer, the Minister for the Environment, Senator Robert Hill, and the Minister for Resources and Energy, Senator Warwick Parer, said in a joint statement: 'Australia will insist that the outcome of current international negotiations on climate change safeguards Australia's particular economic and trade interests.'

Mr John Hannagan, chairman of the Australian Aluminium Council's major policy group, said industry welcomed this statement, 'reinforcing its no-regrets position as its negotiating stand at the forthcoming Geneva talks.' He said: 'This is consistent with the Government's commitment not to support mandatory policy measures which would damage Australia's trade and economic interests. We would also ask the Government for stronger efforts to involve developing countries in the process at the earliest possible opportunity.' (Callick 1996d: 2)

Pearse, in his 2007 book about the players who were blocking climate change action, put public relations consultant John Hannagan and his partner Noel Bushnell (H&B) in context:

Some of Australia's biggest polluters have paid H&B for much of the decade to attend international greenhouse negotiations, write media strategies and press releases, organise conferences undermining Kyoto along with the rationale for emission cuts. Most important of all, polluter money has funded H&B to door-knock the Prime Minister's office (Pearse 2007: 209).

While by this time few government or public documents (other than newspaper reports) are to be found discussing greenhouse science and risk management compared with five years earlier, and the government's *Climate Change* newsletter was now technical and jargon-laden, industry publications continued to sow doubt. The mounting shift in framing can be seen from a July 1996 document called *Greenhouse, not just an environmental issue* produced by the Australian Coal Association. The introduction reflects the 'uncertain' way of publicly communicating climate change that took hold during that period.

The coal industry document asserted that the Australian public was being told by the media and by environmental groups that fossil fuels are to blame for heating the planet, but a more balanced and objective debate is needed.

It also picked apart the 1995–1996 IPCC report to highlight any language signalling uncertainty about human activities affecting the climate, and included contradictory sceptic perspectives. The report highlighted Australia's 'competitive advantage', and called Kyoto target setting unfair. It asked whether, and to what extent, there is a human influence on the greenhouse effect so that action is necessary? In contrast, as the 1980s ended, media and government, and even industry, assumed that this had been thoroughly answered by science.

After 1996, climate change or the greenhouse effect would be commonly presented, in the media and in political rhetoric, as a 'debate' about both the science and about Australia's place in the world in terms of action. The government focus on uncertainty, along with an industry fixation to avoid a carbon tax, guaranteed that the on-ground action would be slim to none despite environment ministers 'rhetoric about 'leading the world' on this and that response initiative. Indeed, there is nothing in the documentary record of this period analogous to the federal and state 'to do' list of the late 1980s and early 1990s. While effective action was minimal, there was a 'vicious attack on the environment movement' during this period (McDonald 2005: 225).

Australia expands coal-fired plants and other sources of emissions

Further evidence on the extent to which both major political parties had by the mid-1990s rejected the early government response framework in favour of a narrative to strengthen the status quo and delay action, was provided in an extraordinary letter. This was sent by a political party, the Australian Democrats, to a climate summit meeting in Geneva in 1996 and reported by Gavin Gilchrist in *The Sydney Morning Herald*:

> In the letter, the Democrats told Mr Chimutengwende [the meeting chairman] that while the Howard Government might claim Australia's greenhouse gas emissions would miss the target of halting their rise by only 3 per cent, in fact Australia's greenhouse emissions were rapidly increasing and almost all of the National Greenhouse Response Strategy remained unimplemented.

> 'The Government is actively encouraging more coal-fired thermal power stations; it does not have the commitment to stand up to the coal industry hence its contradictory attempt to assist marketing coal in the name of greenhouse gas reductions,' the letter says.

'The Government actively encourages more car use by building more freeways and infrastructure to support it rather than improving public transport; it allows far more clearing of native vegetation than is being replaced by tree planting; it increases the number of forests clearfelled for woodchips and it does not adequately encourage development and implementation of renewable technologies.' (Gilchrist 1996a: 2)

A study of how Australia's dominant values on this issues changed during this period concluded that the shift was away from a stance that was global, ethical, risk averse and open to new energy industries. Australia's refusal to ratify the Kyoto Protocol by the late 1990s was portrayed by politicians and the media as normal and logical behaviour to protect jobs and the national interest (McDonald 2005). The new narrative had succeeded by the second half of the 1990s, aided by a small band of familiar sceptics, to persuade the public that 'scientists don't agree', that there is significant uncertainty about the science and action will hurt the economy and, therefore, every family.

'The public interest' drops off agenda

As the story about climate change shifted away from science to economics and costs one cannot overlook the effect of corporatisation and restructuring of government research. The CSIRO is a government science organisation subject to restrictions by the 'employer'—the politicians who fund and thereby direct the research agenda and the organisational structure. In Australia the majority of atmospheric research has been conducted by this semi-independent public entity in partnership with universities and government departments—all of whom are subject to shifting rules about public communication.

A corporate restructure during the 1990s shifted the CSIRO away from public interest research to sponsored research, with a significant impact on the ability to communicate findings. Public interest research, particularly involving the natural environment, was sidelined as CSIRO research was rebadged to serve private enterprise and pay its way with company contracts. The mining and exploration lobby steadily gained influence with federal politicians and 'coal became king' in Australia (Pearse 2009).

The historical evidence suggests that, as the 1990s progressed, some funding levels remained for basic research of climatic processes, but not so for response strategies to reduce energy demand or emissions. Research on energy-efficiency measures and technological solutions that competed with coal went out of favour (Diesendorf 2000; Hamilton 2001). As importantly, public awareness initiatives slowed and eventually stopped after 1992, during the main Keating and Howard government years.

Science focus turned to wealth creation, muzzling inconvenient voices

A parallel outcome was the muffling or muzzling of some of the loudest voices from the scientific community on climate change. Graeme Pearman was one. From his vantage point as chief of the CSIRO Division of Atmospheric Research from 1992–2002 and acting CSIRO Institute director in 1996 he saw 'enormous tension between the mining institute and the environment institute which was considered "too green".' The CSIRO was balkanised. 'The CSIRO Board also became industry dominated,' said Pearman in an interview. 'The paradigm was wealth creation and the role of science is to build wealth ... [this manifested as] denial or that it was politically incorrect to diverge from accountability towards this path.'

Through his role as a science communicator as well as administrator, Pearman learned that the business community had unspoken but adhered-to rules of how it regarded the schism between development and environment or green issues. Even within the scientific community, climate change had become characterised as a special interest green issue and he became characterised as a 'greenie' by the mid-1990s—a far cry from the early scientific mainstreaming of climate science through his efforts and that of other CSIRO scientists in the 1980s and early 1990s. This state of affairs was not helped by the fact that 'people in boardrooms and cabinets don't understand "science speak" on uncertainty', he told me.

By April 2000 Australia had a dedicated sceptical voice on behalf of business with the inaugural meeting of the Lavoisier Group, co-led by the highly influential and active Hugh Morgan and Ray Evans of the Minerals Council of Australia. It was reported at the time that the Lavoisier Group was a spin-off of members from the Australian Business Council who wanted to take a step backwards to dispute the very existence of an anthropogenic greenhouse effect (Taylor 2000).

They were joined by engineers, academics, free market consultant Alan Oxley, and retired government officials—some with considerable clout, such as retired Labor minister Peter Walsh, as well as Brian Tucker, the former head of the CSIRO Division of Atmospheric Research before Pearman—all or most of whom set about contacting politicians (Lavoisier Group 2000).

The express aim of the Lavoisier Group think tank was to debunk, sow uncertainty and otherwise counter the science and policy responses of climate change in the lead-up to possible Kyoto ratification in 2005. In 2004 the group published a sceptic book penned by former Bureau of Meteorology staffer

William Kininmonth that raised desired media attention and debate. Publication of sceptical books at critical policy junctures has been a favoured tactic by critics. In the event, Kyoto was not ratified by Australia at that time.

Internationally, on the other hand, by 2000 the sceptic fossil fuel industry network called the Global Climate Coalition, which had attempted to overwhelm IPCC and political negotiations during the 1990s, was reported to be falling into disunity. A better understanding was emerging amongst key global energy corporations like Shell and Texaco, and car companies like Ford and others, that they should consider the risks and opportunities posed by climate change, not just act as blockers (Windram 2000).

Decline of public attention by mid–1990s

The Australian documentary evidence does not reflect the public reaction to climate change information during the mid-to-late 1990s, as it did with earlier polls. A US poll survey shows, however, that public interest can wane considerably as a story is reframed from a science to a negative political story, as it was during these years (Nisbet & Myers 2007). Some suggestive figures were released by the Australian Bureau of Statistics in 2006, showing public concern for environmental issues in general had declined continuously since 1992 when 75 per cent of Australians expressed concern. By 2004 that figure had dropped to an average 57 per cent.

Since the apparent drop in public interest coincides with the reframed climate change story in the 1990s, this suggests that public awareness or interest about climate change, as well as on other environmental issues, were influenced by similar changed narratives pitting the economy against the environment. Based on the public record, it is unlikely that the loss of public interest came first and influenced politicians and the media to cloud the climate change discussion. The reverse appears to be true.

To better understand not only what happened during the 1990s and since, but also 'how' Australian society was persuaded to forget what it once knew about the mainstream risks and ethical issues attached to climate change, it helps to unpack further the forces that combined to deconstruct Australia's climate change knowledge.

6. Influences on a changed story and the new normal 1990s: values and beliefs

[Prince Charles] has long called on people and politicians to rethink their attitudes to the planet, economic growth and consumption. Recently, however, government policy has become based on the notion that problems such as climate change are best addressed through science and technology—without compromising economic growth and consumerism.

'Greenie Charles worries Labour', *The Australian*, 11 July 2009

Environmental questions are inextricably intertwined with economic issues and, at base, are concerned with values rather than so-called 'scientific facts'.

David Mercer, *A question of balance—natural resources conflict issues in Australia*, 1991

To understand what happened in Australia in the 1990s, with its dramatic change of public narrative on climate change, it helps to explore the ideas that came with the opinion leaders who reset the narrative. This is important not least because the same influential ideas are back in the political driver seat with renewed vigour, having never really left since the mid-1990s.

In November 2013, just before a super typhoon, packing 315-kilometre-per-hour winds, ravaged the central Philippines with many thousands dead, cities flattened and the resulting hunger and desperation,[1] former Prime Minister John Howard told the world what he thought of climate science and the need for evidence-based policymaking. Howard led the Liberal Party and the country from 1996 to 2007. He was speaking at a meeting of the Global Warming Policy Foundation (GWPF), a UK gathering of climate 'agnostics' and 'sceptics'.

He is reported to have said that 'the high tide of public support for overzealous action on global warming has passed' and that 'I instinctively feel that some of the claims are exaggerated' while also repeating the common sceptic thesis that climate scientists are just after public money by scaring the public about upcoming catastrophe (Hall 2013).

1　Scientists and journalists who follow the climate change story commented that no single typhoon or cyclone along with storm surge can be attributed to global warming and sea-level rise. However 'scientists do have more confidence about what's likely to come in the decades ahead if we can't curb global warming: stronger tropical storms … the typhoon's strength could well be a sign of catastrophes to come' (Walsh 2013).

In the breaking news story on this event, Fairfax London correspondent Nick Miller reported that Howard called climate change predictions 'alarmist' and linked his own position to economic management and growth, particularly in developing countries—which happens to benefit quarry Australia. Miller also revealed that Howard admitted the only book he had read on climate change was a 2008 tract by (Lord) Nigel Lawson, an avowed climate sceptic and GWPF founder (Miller 2013).

The Fairfax media had a bit of editorial cartoon fun with Howard's remarks linking the new administration of Prime Minister Tony Abbott to Howard's legacy and ideas:

'Oh Lord protect us from the zealotry of climate scientists …' begins one cartoon by Wilcox in *The Canberra Times* of 10 November showing Howard, Abbott and environment minister Greg Hunt piously in the church pews.

The week before, Hunt declared that he would not attend the latest international climate change talks because he needed to be at home for the first parliamentary sitting whose top legislative agenda item was to undo the carbon price established by the previous Labor administration. The Abbott government in its first month in office also dismantled the country's Climate Commission of scientific advisers and proposed to defund a public finance corporation that lent to renewable energy projects and which, incidentally, was profitable, according to Hansard records.

Here was a timely reminder that Howard's science-free beliefs guided the public face of Australia's policy positions on climate change for more than a decade before he was defeated in 2007. In 2013, this set of beliefs was resurfacing strongly.

There's plenty of evidence, as I have shown in the previous chapter, that outside the halls of Canberra a wider campaign of confusion and doubt supported Howard's instincts. That campaign has continued since Howard was in government. Look no further than 2009 in the lead-up to the '15th conference of the parties to the UN Framework Convention on Climate Change (UNFCCC)' held in Copenhagen. Work by an international consortium of investigative journalists, including in Australia members of *The Sydney Morning Herald* staff, unveiled the continuing and highly funded public relations campaigns by multinational coal, aluminium, and electricity industries and, at times, related labour unions, to influence and deflect government action ('The global climate change lobby'). The lobbying story, even some of the players, were the same as previously documented and discussed looking back at the 1990s. The tactics were also familiar.

Familiar themes invoked a mental pathway where jobs and family 'us' would lose out should action on climate change involve any change from the energy system status quo; for example, employers would go offshore if their needs for cheap power were not met. This narrative has made the economic interests of multinational companies synonymous with the national interest and rhetorically with every family's interests.

With the link between the increasing risk of extreme weather events and global warming and climate change having been laid out time and again by the Intergovernmental Panel on Climate Change (IPCC) assessments, national academies of science and even some business sectors, such as the insurance industry, the ongoing denial that there might be a problem begs the question: have influential opinion-leaders behaved irrationally during the past 20-plus years and do they continue to do so? How can they deny the risk and subvert action faced with the overwhelming scientific evidence of outcomes that are more evident on the ground year by year?

While many observers believe that politicians and their advisers are unable to think past the current or next election cycle, thus being unable to respond to a long-term issue like climate change, there is more going on. Underlying the changed narrative are values and beliefs held by politicians and other influential opinion leaders that make their responses internally consistent, if not always 'rational' or understandable to the casual observer. These values and ideas dominated society at large during the 1990s and 2000s and thus came to seem 'normal'.

Values guiding denial

The denial of environmental reality, what is happening in the natural world around a civilisation, is not a new phenomenon for the human species. Biogeographer Jared Diamond looked extensively at the collapse of previous civilisations and says that, more often than not, it has been the values and beliefs of societal elites more than any empirical on-ground evidence (frequently of changing climatic conditions) that has determined the fates of these civilisations (Diamond 2005).

The proposition that values guide much environmental policy development and communication takes the story beyond a saga of corporate self-interest from potential losers in energy sector reform. Instead we see that facing the climate sciences during the 1990s and since have been deep-seated ideologies that came to exercise a comprehensive grip on Australian society. The term hegemony would fit the cultural consensus that prevailed in Australia post 1996. (Italian political scientist Antonio Gramsci's influential theory of hegemony postulates that in advanced industrial societies, one group or class can rule through dominating everyday ideas and practices, and this is done through information—for example, mass media and public relations, schooling, popular culture, and consumerism.)

Dominant ideas and values have built on a modern 'no limits' view of human capability versus the natural world. There are parallel beliefs in growth and progress as guiding principles in the organisation of society; plus related beliefs in the power of technology to fix all problems (a technocratic view of the world that has gone into overdrive since the ascendance of the internet in the early 1990s, spread by the many companies that have capitalised on its opportunities).

Also tapped are older, deeply embedded beliefs in human exceptionalism from the rest of the natural world. These unspoken beliefs are shared by opinion-makers and much of the public, and came together in the 1990s with a particular form of free market capitalism (economic rationalism) and its related economic assumptions. Together they reshaped the political and media responses to the scientific messages, and came to define the dominant narrative for the general public: what we all should believe.

Pioneering beliefs and techno fix

Like other 'pioneering' Western democracies, such as the United States and Canada, Australia has long-standing beliefs that natural resources are there to be exploited and developed. Australia also enjoys a recent history that seems to prove (European) humans can always come up with a techno-fix solution. With this world view, people can believe ecosystems, plants, animals and natural resources, even nature itself, do not exist independently only as a store of matter and energy waiting for human transformation. 'The most important natural relationship taken for granted by [this belief system] is a hierarchy in which humans (and in particular human minds) dominate everything else' (Dryzek 1997: 50).

This is a hallmark of a secular techno-fix world view that has dominated since World War II. While not necessarily denying that natural systems exist in their own right, proponents of this view often subscribe to a 'wise use' philosophy to argue that not to exploit the planet is 'wasteful' and that adverse environmental effects are overstated by special interests such as 'greenies', who do not have jobs and the national interest at heart (Beder 2000; Dryzek 1997). The corollary idea is that humans will always find a technological solution to any environmental problem anyway and that 'no-limits' development is the natural order of human endeavour.

Traditional adherents to this philosophy have been the big mining and timber production companies, farmer and pastoralist organisations, professional engineering bodies and the like (supported by development-oriented State and Commonwealth governments) ... and rhetorical links with 'progress', 'national interest', 'wealth/job creation', 'development', 'growth', 'defence' and so on have frequently been made. (Mercer 1991: 41)

Monash University geographer David Mercer wrote the above from a historical perspective in 1991. At that time, in response to greenhouse risks outlined by scientists, a set of ethical values, based on the public interest and global responsibility, were guiding Commonwealth and most state policies, and were being communicated to the public. What happened in the 1990s is that a more traditional 'no-limits' view of humans versus nature and the sectors listed by Mercer appeared to reassert themselves, boosted by the economic ideas of neo-liberalism.

In Australia, regardless of the nominal differences between the major political parties, values have been shaped across the board by the colonial and pioneering experience based on vastly modifying the natural environment for agriculture and later quarrying for mineral wealth. This belief that Australia's economic and cultural destiny is to sell natural resources and be a (well-paid) quarry for the rest of the world sits comfortably with the 'no limits' and techno-fix narrative.

The problem that evolved was what to do with the conflicting idea of human-generated climate change due to burning fossil fuels when Australia became a world-leading coal exporter and domestic use was also running at full throttle. Denial and argument might then come naturally, accompanied by occasional rhetoric about bringing emissions down.

A wishful Australian techno-fix solution to this mental conflict has been 'clean coal' or carbon capture and storage (CCS) in the coal-burning cycle. There has been little scientific evidence that CCS will do what it is supposed to do, but government rhetoric often proclaims otherwise. So far, the focus on clean coal has been a matter of belief rather than evidence. Atmospheric and earth scientists have offered no such techno fix in any timeframe that matters to current emission statistics.

The environmental movement itself often subscribes to a techno-fix world view. A 2005 interview study with 25 of the top leaders of US environmental organisations found a biological systems approach lacking in environmental policy and concluded:

Thinking of the environment as a 'thing' has had enormous implications for how environmentalists conduct their politics ... (which) hasn't changed in 40 years. First, define a problem (e.g. global warming) as environmental. Second craft a technical remedy (e.g., cap and trade [in Australia, emissions trading]). Third, sell the technical proposal … through a variety of tactics such as lobbying, third party allies, research reports, advertising and public relations. (Shellenberger & Nordhaus 2005: 4)

These authors argue that even a reliance on individual choice for LED lightbulbs or hybrid cars reflects this same techno-fix mindset—if only we have enough technical solutions understood by the public, this problem will be solved. They also identify reasons why environmentalist organisations could be as easily marginalised as they were in Australia in the 1990s. They suggest a narrow tactical focus and lack of effective coalitions with other societal interests (e.g., labour unions, or animal welfare organisations, or church groups), have not helped. The internal 'group think' encourages the broader society to consider the environment to be separate from mainstream concerns.

Progress myth and human exceptionalism: Beliefs that trump science

In Western democracies with Western Christian values, it is telling that the following view is hardly considered unusual. In a 2004 opinion piece, syndicated columnist Angela Shanahan, who doesn't hide that she is a practising Christian, slammed those citizens who would protect Australian wildlife against lethal state government programs. She called such people 'extreme greenies', with 'unreal', 'Mickey Mouse', 'anthropomorphic' world views that deny there is 'such a thing as a hierarchy of living things' (Shanahan 2004).

A mythology that humans are exceptional, top of the hierarchy, and not subject to the 'laws of nature' governing other animals, underlies much of Western thinking and stems from Christian teachings. Arm-in-arm with exceptionalism is the always forward-looking, linear mythology of 'progress' and human betterment. Historical philosopher Ronald Wright wrote about belief in progress as cultural myth: 'Myth is an arrangement of the past, whether real or imagined, in patterns that reinforce a culture's deepest values and aspirations … Myths are so fraught with meaning that we live and die by them' (Wright 2004 4).

Historian Lynn White Jr, in his ground-breaking essay 'The historical roots of our ecologic crisis', which was published in *Science* in 1967, argues that the fundamental religious myth of humans as exceptional is both most pervasive and most internalised in Western culture. 'What people do about their ecology depends on what they think about themselves in relation to the things around them. Human ecology is deeply conditioned by beliefs about our nature and destiny—that is, by religion' (White 1967: 1205).[2]

White wrote that, in its Western form, as distinct from the more contemplative Eastern Orthodox church, Christianity is the most anthropocentric (human-focused) religion the world has seen. It has a world view that denies the existence of any spiritual qualities (the soul and similar concepts) in other species. From there extends an assumed exceptionalism to the laws and needs of the natural world that brings many people into conflict with environmental science or ecological knowledge. 'Christianity … not only established a dualism of man and nature, but also insisted that it is God's will that man exploit nature for his proper ends' (White 1967: 1205).

White also identifies the 'progress myth', saying Western (Judeo-Christian) cultural activities are dominated by an implicit faith in perpetual progress, which was unknown either to Greco-Roman antiquity or to the Orient. In fact, Marxism, which is superficially anti-religious, is provocatively called a Judeo-Christian heresy in this analysis, due to its beliefs, along with capitalism, in the guiding myth of perpetual progress. Forty years ago, when environmental studies were starting in earnest, White predicted that ecological crises will worsen as long as people, including many scientists, retain these unexamined basic assumptions and myths.

It's easy to overlook that Western science and technology were born from the desire to understand God's works and that, when this desire merged with the Industrial Revolution in the 19th century, it allowed ever greater exploitation of natural resources—believed to have been put there by God for man's benefit. Science historian and physicist Spencer Weart, while giving a detailed account of the scientific discovery of global warming, concludes that beliefs, including religious beliefs, will guide future climate change because they guide how we deal with our environment in an age when humans are altering planetary systems (Weart 2004).

There is an intellectual kinship between Christian beliefs in the exceptional nature of humans on the planet and the more secular progress, 'no limits', views of human activity and impact. Armed with these various beliefs of limitless possibility, it may seem rational to dismiss a precautionary, risk-management response to scientific assessment of climate change—and to the population growth that implicitly drives more greenhouse gas emissions. Leading agenda-

---

2   White wrote in 1967, in the context of human impact on the planet, 'our present combustion of fossil fuels threatens to change the chemistry of the globe's atmosphere as a whole with consequences which we are only beginning to guess' (White 1967: 1203). Another piece of evidence for the general knowledge about climate change that existed decades ago.

setters in politics, the media, and business can appear to believe (demonstrated by what they say and do) that what is happening to the natural world will not affect human culture. For similar reasoning, such beliefs could encourage a person to deny or dispute that there are global warming impacts and links to extreme weather events.

The websites of organisations like the Lavoisier Group, or the environmental arm of the Institute of Public Affairs (IPA), and other free market Australian think tanks reveal such contrarian thinking. In a revealing 2004 article on the activities of the anti-climate science Lavoisier Group, *The Age* journalist Melissa Fyfe characterised those at a meeting of 50 men (only one woman) in Melbourne as follows: 'Some of them were scientists. But many were engineers and retired captains of industry. Presiding was Hugh Morgan, president of the Business Council of Australia and former Western Mining boss. The master of ceremonies was retired Labor politician Peter Walsh' (Fyfe 2004: 1). To the extent that 'no limits' beliefs and values seem natural to these institutions and professions, their opposition to accepting climate change science is more understandable.

Ecological limits: Understood since the 1960s and accepted until the late 1980s

Opposing the 'no limits' view is a belief in ecological limits and limits to the carrying capacity of the planet, and indeed the realisation and scientific evidence (including DNA analysis) that humans are one among many evolved species and, in many ways, not so exceptional.

This understanding was well developed during the first wave of global environmentalism starting in the 1960s. The understanding of limits was popularised by the global think tank Club of Rome reports in the 1970s and the works of biologists Paul and Anne Ehrlich, and Lester Brown of the Worldwatch Institute, and many biological and other scientists in the past 30–40 years. The understanding of limits still informed the climate change narrative in the late 1980s, as the documentary evidence shows.

E.O. Wilson, Harvard ecologist and a leading theorist on the interaction of humans with the natural world, has observed that human beliefs about our prospects on the planet now fall into two categories: human exceptionalism and environmentalism. Environmentalism here defines humans as a biological species tightly dependent on the natural world, compared to feeling freed from the biophysical constraints of the planet by transcendent intelligence and technological prowess. Humans' conflicted views on their own skyrocketing population numbers is a related example of exceptionalist thinking (Wilson 2005).

Wilson noted that the short-termism that marks not only politicians, but also the species in general when faced with anything other than self, family, or tribe, may have had an evolutionary advantage over the two million years that modern humans evolved, where life was mostly precarious, short and unpredictable. Evolutionary biologists have noted that modern humans bring a Palaeolithic hardwire to runaway technical success.

It can be argued that the late 1980s upsurge in multinational, energy-intensive industries, such as aluminium and mining in Australia, encouraged a strong return to the traditional, no-limits and exceptional way of viewing humans versus the natural environment and Australia versus the rest of the world. Certainly the mental gymnastics involved with promoting all-out coal production while giving lip service to climate change action would indicate that the thinking has been 'rising emissions will not affect us'. These beliefs merged smoothly with a defining economic system that gathered steam and reached its height in the 1990s led by a new breed of economist.

Rise of the economists and fall of the public sector

> The ideas of economists and political philosophers, both when they are right and when they are wrong, are more powerful than is commonly understood. Indeed the world is ruled by little else. Practical men, who believe themselves to be quite exempt from any intellectual influence, are usually the slaves of some defunct economist. Madmen in authority, who hear voices in the air, are distilling their frenzy from some academic scribbler of a few years back. I am sure that the power of vested interests is vastly exaggerated compared with the gradual encroachment of ideas. (John Maynard Keynes, 1936: 383)

In the last 20 years, the public record indicates that beliefs in the wisdom of technology and underlying exceptionalism values, married to economic dogma, have exerted far more influence on policy and communication than scientific evidence. A federal and state policy adviser during the 1990s observed in an interview: 'The biggest barrier (to effective greenhouse action) is the intellectual mindset of economists and their belief systems about whether government should be a player in changing the economy'.

Frank Muller called those who advised government based on these beliefs 'econocrats'. In econocrat thinking, following neo-classical economics, governments should not regulate or interfere in markets because all efficiencies are already in the system. So, requiring energy efficiency measures, for example, constitutes unwarranted interference in the relevant industries and markets.

Comparing his Washington D.C. experience during the administration of President Bill Clinton in the mid-2000s to an ongoing situation in Australia, Muller said:

> In Canberra we have not had the countervailing science influence to the econocrats. There are not people presenting the story to the decision-makers that 'we can do something' about (climate change) right now. So the denial in government policy circles is real … their experts are telling them they don't have to deal with it right now.

The free market ideas that inspired not only Australian policymakers in the 1990s, but the English-speaking Western democracies in general, were a revision of neo-classical liberal economic theories, known as 'economic rationalism' in Australia. (Thatcherism, Reaganomics, and the Washington Consensus are other labels on the neo-liberal theme). This set of economic assumptions, beliefs, and values was gathering steam in the 1980s, and came to govern Australian public policy and national conversations as the 1990s progressed—hegemony in action.

Trading an ethical, responsible world view for economic self-interest

The previous 'ethical' understanding of climate change risk (stressing the public interest, intergenerational equity, inter-country responsibilities, etc.), was replaced by a world view promoting economic self-interest as the dominant value. The public was encouraged to focus on its consumer rights while citizen responsibility took a back seat.

This shift away from moral values and responsibilities was highlighted in a surprising press release that appeared in November 2006. Heralded as a world-first event, 16 Australian religious denominations spanning all faiths, issued a joint statement regarding the need to value the planet and life. Skirting national and religious exceptionalism, concern about climate change was described as a core matter of faith and morality.

The release said that politicians will be held accountable to 'do something' about addressing climate change. 'It's not just about the price of coal, or about whether we can't do anything … it's absolutely important that such a large issue … is reflected in our own moral beliefs, whatever faith they are, when you're confronted by the nature of this kind of challenge' (Crittenden 2006). An umbrella movement called green Christianity does, as in some other religions, view the natural world as one of kinship and humanity's role as one of stewardship, rather than a relationship of domination and exploitation.

Following the 2006 statement, the Australian Religious Response to Climate Change (ARRCC) network was established in 2008 to provide educational and campaign resources to faith communities to respond to climate change and lead by example with environmentally sustainable practices. In 2014 members of the ARRCC were being arrested in protests against a proposed coalmine in in the Leard State Forest in northern New South Wales.

The religious challenge, similar to that of others interested in safeguarding the natural environment from carbon pollution and other negative changes, is reacting in part to the demise of the traditional concept of 'public interest' under economic rationalism and the loss of ethically based policies for climate change action.

A change in the perception of what constitutes the public interest and the validity of government action on behalf of the public versus private sector interests was closely connected with the ideas popularised by a new generation of economists. Some 50,000 graduated between 1947 and 1986 according to social scientist Michael Pusey, who published a definitive study of the Canberra bureaucracy in 1991.

These graduates brought both neo-liberal economic ideology and a narrow technocratic training to the policy arena. This was in comparison to a broadly humanistic training, 'liberal' in another sense, which was common before World War II. Contemporary economic specialisation: 'almost invariably excluded any broadening study of philosophy, sociology or the history of ideas' (Pusey 1991: 172). (This is true also for the study of other technical and scientific specialisations by the 1980s and 1990s.)

A neo-liberal or neo-classical, economic rationalist world view assumes that individuals and businesses, given the freedom to choose, will rationally choose to maximise gain and profit, and everything else flows from that. The new narrative, amplified by the media as politicians and think tanks talked this language, stopped discussing risk, equity, and the public interest in regard to climate change, and instead focused on costs and the dollar bottom line to consumers and business. This was framed for public consumption as the proper concern for jobs, family, and national interest—that is the mainstream interest.

In practice, economic rationalist thought and policies (including national competition policy) quickly dispensed with ambitious state plans for energy efficiency measures, renewable energy and fuel substitution that were developed before 1992. The dominant narrative changed to assign unattractive costs to efficiency and renewables (compared to conventional generation of 'cheap' electricity) on the theory that the energy market is already at maximum efficiency.

Of course the computer models that underpinned these changes are only as useful as their baseline assumptions. Economic rationalist models reached their cost assessments while discounting some factors. They did not, for example, factor in the cost to society of large government subsidies supporting mining and coal-fired electricity. They also ignored 'externalities'—the costs of production that can be shunted outside the corporation or producer, including the cost of environmental consequences such as emission of greenhouse gases. Economic rationalist modelling also appeared not to give equal weight to any potential *benefits* arising from renewables or efficiency over time (Industry Commission 1991).

The lessons of economics

Universities have been the incubators of changing economic theory, including economic rationalism in Australia, during the past 40 years. The lessons taught as basic economics have influenced a generation of bureaucrats, politicians, and policy advisers.[3] Relatively little attention, however, has been paid to what is actually taught to impressionable young minds.

American journalist Christopher Hayes spent an academic quarter auditing 'Principles of macroeconomics' at the University of Chicago and provided a snapshot of the material that was taught in the course. Efficiency is the defining value of the Chicago School of Economics (home base for the late Milton Friedman's version of neo-classical or neo-liberal economics, which resonated in Australia), and is still the basis for instruction. 'Too much' government (regulation or public ownership of assets or services) causes inefficient economies in the overall quest for capital growth (Hayes 2006).

The international capital market is seen as the primary regulator of a society, and that view continues to prevail despite the global financial crisis of 2008. The conversion of natural and human resources into capital growth will raise everyone's standard of living, domestically and globally, without government regulation. With 'no limits' and 'endless growth' forming an embedded part of this thinking, questions of sustainability have not applied.

---

3   Jones (2002) noted that The Australian National University School of Economics was prominent in producing the home-grown neo-liberal theorists and free trade economists who advised Labor and Coalition governments from the 1980s on.

This version of economics is called neo-classical because the injunction to 'specialise and trade' harks back to Scottish political economist and philosopher Adam Smith's 18th century insights about 'comparative advantage' and what creates the 'wealth of nations'. John Howard regularly invoked 'our comparative advantage' of exporting fossil fuels when defending Australia's limited actions to lower emissions.

In this economic universe, normative models are transformed into reality. Translated that means: arguments about the way the world *should be* are converted into assertions about how the world actually is, without the need for empirical data or evidence (in 'econ speak', converting normative arguments to positive statements). Thus, 'people cannot disagree with neo-classical economics, they can only fail to understand it' (Hayes 2006: 28).

This version of economics presents itself as a value-neutral description of how the world is: therefore students do not perceive they are learning an ideology. (In a similar fashion, Keynesian regulated capitalism enjoyed a consensus of 'this is the way the world is' before and after World War II—until the early 1970s.)

Hayes learned that theory is demonstrated through economic modelling—simple supply and demand at the level of tertiary education—and is taught regardless of real-world empirical studies that indicate the facts can be otherwise. Hayes also witnessed the use of contrarian, sceptical positioning as a tactic to skewer opposing points of view, institutions or consensus. This is familiar territory in the successful deployment of contrarian debate in the public discussion on climate change.

The defining characteristics of present-day economics

At The Australian National University Centre for Resource and Environmental Studies, economist Jack Pezzey told students at a 2006 seminar that 'economics is a mind-set' and that several relevant defining characteristics of the discipline of economics as we now accept it are (my italics):

- Anthropocentricity—there is no value unless it is related to human notions of value; thus environment is seen as an 'amenity' or 'input'
- consumer sovereignty—the neo-classical view is that economics *reflects* people's preferences, rather than shaping them (*even though people's preferences are always shaped by the culture around them*).
- non-satiation—the notion that people or firms always prefer more 'well being' (*translated as profit*), rather than less

- aggregation—the focus on average or total variables, not their distribution, which is the province of politics
- finite trade-offs—nothing is beyond price
- discounting—means that future costs and benefits are always worth less than today's. (*Thus, economically, climate change happening in 30, 50, or 100 years time is uninteresting*).

The gold standard of neo-classical economics is the marginal cost of supply, which has led to the demand for 'no regrets' solutions to climate change, meaning there should be no increased cost to the economy or individuals. 'No regrets' has since become bureaucratic jargon applied to various transactions, usually with minimal understanding by an outsider of what the term means.

There are alternative schools of economic thought; for example, ecological economics rejects many of the key concepts of neo-classical economics and its 'no-limits' assumptions, and can pose deeper philosophical challenges to the anthropocentric value system.

Where did this thinking come from?

English historian Richard Cockett has traced the current thinking of neo-classical free enterprise to the so-called Austrian School at the London School of Economics. Led by Friedrich Hayek, Karl Popper and Lionel Robbins, this school began criticising the then prevalent Keynesian economics of regulated and 'welfare capitalism', starting in the 1940s.

Friedman and his colleagues of the Chicago School at the University of Chicago were influenced by the English economists and proposed similar radical remedies starting in the 1950s—to end regulation of economies and to minimise the state/public sector. Friedman was later to have direct influence in Australia through his thoughts on globalisation and capital markets, which were taken up by both major Australian political parties, and as a speaker invited by think tanks (Klein 2007). These theories and thinkers started their global influence with biannual international meetings at Mont Pelerin in Switzerland; these meetings were the origins of the Mont Pelerin Society that provides an international network to this day (Murray & Pacheco 2000).

A myth to live by

Economic rationalist theorists decreased the role of government support and industry protection while opening Australia more to international financial markets: moves that were supported by both major political parties. The enthusiasm for 'free trade' and 'free markets', and the underlying assumptions of neo-classical economics, was accepted as early as the 1970s in Gough Whitlam's Labor Party and was reconciled, more or less, by successive governments with Australia's industrial relations framework, environmental regulation and other forms of social good up until the 1990s.

While recent history shows that economic rationalist restructuring of the economy has combined with international market demand in Asia generally and China specifically to produce a period of increased national affluence, the ideology has also attracted critics of its underlying assumptions in a longer term context. Political economist Evan Jones characterises the evolution of economic rationalism into the dominant way of looking at the world as opportunistic and neither coherent nor logical, but rather a convenient 'myth to live by' driven by belief rather than a background of empirical evidence. He suggests that 'the universalism of the solutions is a clear indication of their religious character: ideology has rushed in to fill the vacuum left by the poverty of analysis' (Jones 2002: 57).

Nobel Prize-winning US economist Joseph Stiglitz reportedly voiced a similar critique saying neo-liberal theories can best be understood as another belief and values system, rather than being based on empirical facts. The dominance of the neoclassical model is 'a triumph of ideology over science' (Hayes 2006: 27).

Politics as battle of ideas

In the end, politics is a battle of ideas and a battle of commitment.

John Howard, 2002

Economic rationalism finds its historical place within a 'battle of ideas' that has been waged simultaneously with the environmental movement's rise in the late 1960s and 1970s. Parts of the ideological underpinnings stem from 19th and 20th century 'wise use' philosophy. Wise use and its variants developed in pioneer countries like the United States and Australia where influential elites were also against government regulation in general and particularly if applied on behalf of the natural environment (Ehrlich & Ehrlich 1998). In this value system rights include the individual's right to derive wealth from nature. Wise-use thus also links to previously noted religious thinking of human domination over nature. These beliefs easily convert to anti-environmentalism and exceptionalist arguments.

Free enterprise, individual property rights and 'wise use' were philosophical cornerstones of Reaganomics or supply-side economics in the United States, Thatcherism in the United Kingdom, and economic rationalism in Australia. Deregulation and competition policy have been other cornerstones leading to privatisation of public assets, including electricity generators, as well as an all-out commitment to 'free trade' by both major Australian political parties ('Labor in power' 2010; Sturgess & Torrens 2009).

Before looking more closely at how competition policy affected the climate change story, it's worth asking how these ideas, building on underlying beliefs and values, became part of the communication mainstream and were embedded in Australian institutions and in society at large.

Think tanks supplied the ammunition

Right-wing or pro-market think tanks have generated much of the rhetorical ammunition for neo-liberal policy. They have enjoyed considerable influence in Western English-speaking countries in the past 30 years—specifically in the United States, United Kingdom, Canada and Australia,[4] with the understanding that changing the dominant narrative was the pathway to success.

'Like other movements, the main impact of the radical neo-liberals was not direct policy influence but broader discursive shifts ... demonising and disorganising opponents of neo-liberalism' (Cahill 2004: 24). Tactical advice, high-profile speakers and organisational liaison between countries has been common. 'Because of this, the ideas of Friedrich Hayek, Public Choice Theory, Milton Friedman and developments in neo-liberal theory and neo-liberal policy alternatives have been disseminated in Australia' (Cahill 2004: 8–9).

Think tanks are non-government policy centres funded by corporations, private money and the taxpayer. They have overtaken more informal networks (the club, the school or university, the corporation) as vehicles for ideological battle. They have adopted the trappings of academic and scientific research—the conference, lecture, and journal. Their thinkers are able to be far more definite and far less constrained by peer review on issues like climate change than the scientific convention dictate. They publish the writings of sceptical scientists,

---

4   Cahill, following the seminal work of Cockett in the mid-1990s, documents how overseas think tanks influenced neo-liberal thinking in Australia and the establishment of Australian think tanks: 'Radical neo-liberal organisations such as America's Heritage Foundation, Britain's Institute for Economic Affairs (IEA) and international networks such as the Mont Pelerin Society and the Atlas Foundation served as examples for the Australian movement to emulate' (2004: 8).

which are seldom peer reviewed or their authors questioned on their relevant expertise (Cahill & Beder 2005; Jacques, Dunlap & Freeman 2008; Murray & Pacheco 2000).

The fossil fuel industry-supported IPA is the oldest and probably best known of the Australian free market, neo-liberal think tanks and has long been associated with the Liberal Party in executive roles. IPA Executive Director John Roskam has worked for both the government of Jeff Kennett in Victoria and the Howard federal government in the 1990s. Policy positions advocate privatisation and deregulation; limiting the role of unions and non-government organisations on policy and contesting the science and policy initiatives involved with environmental issues such as climate change.

Its environmental arm has been active in communication and, during the 1990s and since, it has attacked climate science in books and articles by sceptic academics, such as geologists Ian Plimer and Bob Carter, former bureau meteorologist William Kininmonth and former CSIRO division head Brian Tucker.

'Doubting Australia: the roots of Australia's climate denial' is a document compiled by Australian non-government climate action groups with assistance from sourcewatch (a collaborative watchdog wiki directed by the US Center for Media and Democracy that produces documented information on policy-oriented think tanks and their activities). It credits the IPA as being 'at the heart of climate denial in Australia' since the early 1990s when it started inviting US sceptics to this country. Links to US and UK think tanks (including wise-use philosophers) strengthened after 1996 with well-publicised sceptical events (including with (Lord) Nigel Lawson's foundation where Howard spoke in November 2013) ('Doubting Australia' 2010).

According to 'Doubting Australia' (which offers 118 links to source documents for its exposé of think tank connections to climate change denial in Australia since the mid-1990s) the IPA's support for sceptical geologist and mining company director Plimer and his 2009 book *Heaven and Earth*, which argues that global warming is not happening, has been very effective communication. The book reportedly sold more than 40,000 copies in Australia and it was supported not only by the Murdoch press, but reportedly defended by the Australian Broadcasting Corporation (ABC) as a legitimate balancing voice on climate science, despite Plimer's lack of scientific credentials in that area and the factual mistakes that critics found in his book.

Plimer's book impressed then Opposition front bencher Tony Abbott, who was quoted in a 2009 ABC *Four Corners* program on the proposed emissions trading scheme in the following terms: 'I think that in response to the IPCC alarmist—ah in inverted commas—view, there've been quite a lot of other reputable scientific voices. Now not everyone agrees with Ian Plimer's position but he is a highly credible scientist and he has written what seems like a very well argued book refuting most of the claims of the climate catastrophists.' The program provides direct testimony from Liberal and National Party politicians, now in power, of their divided or negative views on climate change science and action (ABC 2009).

Plimer toured Australia for his book with entertaining but science-free UK climate sceptic (Lord) Christopher Monckton, claiming an emissions trading scheme would damage the mining industry with irrational demands. He also has signed international documents attacking the premises of mainstream climate science. This included an international petition by retired geologists, meteorologists (including Kininmonth) and others to Canadian Prime Minister Stephen Harpur urging him to dismiss emission reduction proposals as the '"catastrophist" fabrication of environmental groups' (*Providing insight into climate change*).

In the run-up to the 2013 federal election, Abbott was publicly associating with the IPA, most notably responding to its radical policy wish list with a rhetorical 'big fat yes'. The list started with 'repeal the carbon tax', abolish the Department of Climate Change and abolish the 'clean energy fund'. At the time he also attended a well- publicised IPA birthday dinner with much mutual backslapping with mining billionaire Gina Rinehart and News Limited owner Rupert Murdoch, whose newspapers pulled out all the stops to elect Abbott and his party.

The Tasman Institute, established by former Monash University economics lecturer Michael Porter in 1990, like the older IPA, adheres to economic rationalist theories and positions, with direct and indirect relevance to climate change policy development. The Institute's 1995 annual review revealed a 'who's who' of Australian resource industries among the 21 corporate members and its flagship project at that time was called 'Markets and the environment'. Its focus was issues affecting investment in Australia's resource-based industries (Maddox 2005). The Tasman Institute was amongst the first to criticise Australia's proposed climate change response strategy (*Economics and the environment* 1990).

Think tank influence was overt in the case of Project Victoria in the early 1990s with a market-forces deregulation blueprint produced by the Tasman Institute and the IPA accompanying the new Liberal government under Kennett in 1992, which put a stop to ambitious state plans for energy efficiency in domestic and commercial sectors and the state electricity commission plans to decrease consumer demand (Cahill & Beder 2005).

Other influential Australian think tanks with similar economic orientation are the Committee for Economic Development of Australia (CEDA), the Centre for Independent Studies (CIS) and the Business Council of Australia (BCA) (Murray & Pacheco 2000). The CIS started in 1976 in the garage of then high school maths teacher Greg Lindsay, going from rags to riches thanks to corporate sponsorship from Western Mining's Hugh Morgan among others. CIS linked economic rationalism with social conservatism under Christian and family values icons. Researchers have called it a favoured brain trust for not only Rupert Murdoch, but also John Howard during his tenure as prime minister (Maddox 2005).

During the 1990s, the BCA's president Hugh Morgan and his associate from the mining industry, Ray Evans, featured prominently in reports on the development of not only the Tasman Institute and CIS, but also other pro-market, socially conservative, Christian values and scientifically contrarian think tanks in Australia (Hamilton 2006; Maddox 2005). That list includes the Lavoisier Group, established in 2000 specifically to counteract climate change science, with Morgan as the first president and Evans as secretary (Lavoisier Group).

Speaking at the 50th anniversary of the conservative journal *Quadrant*, Howard was candid about the 'war of ideas' that must be fought. He reportedly named Ronald Reagan, Margaret Thatcher, and Pope John Paul II as the 'towering figures' of the late 20th century for their moral clarity and ideological opposition to all collectivist thinking as 'stultifying orthodoxies and dangerous utopias.' He spoke of the historical battle of ideas for Western civilisation and 'the essential connection of personal, political and economic freedom' (Schubert 2006).

With think tanks delivering a steady stream of reports and ideas to politicians and to like-minded media editors, particularly in the Murdoch press, economic rationalist beliefs couched as 'the economy' came to dominate public discussion and thinking (which is a good working example of hegemony) by the mid-1990s and particularly thereafter. But already with Labor under the Hawke government at the turn of the decade, a dominant pattern had been evolving around market economic thinking. 'Ministers and their top staff see the world very much as male age-mates through a shared and restricted formative training in economics' (Pusey 1991: 8).

The 'public interest' becomes the corporate interest

Championing the neo-liberal agenda 'Australia's foreign-owned media, the New Right think tanks and research centres … have had an enormous success in penetrating the Canberra apparatus, and [also] international economic organisations such as the World Bank and the OECD' (Pusey 1991: 13). This economic set of ideas was extended to the administration of semi-independent organisations, such as the CSIRO, as they were corporatised and directed to serve industry. The economic rationalist world view increasingly painted the public interest as indistinguishable from that of large extractive industries.

Federal and state bureaucracies realigned their thinking according to the new economic orthodoxy. In terms of climate change response, they came to downplay the benefits of the easiest response—energy efficiency strategies. Since the economic rationalist view was that markets are, by definition, ultimately efficient, the conclusion was that mandating, for example, efficient commercial or industrial facilities or household appliances must add costs and 'intervene' in markets unnecessarily. The net effect was 'industry capture', which made progress on greenhouse policy extremely difficult (Hamilton 2001: 33).

Environmental consultant and RMIT academic Alan Pears worked in the area of energy conservation and efficiency in Victoria from the 1980s on. He saw the changes that economic ideology wrought and he noted that, in 1994, the federal Treasury Department of Paul Keating's Labor government released a document bowing to a 'perfect market structure' and a focus on unfettered competition.

The capture of the bureaucracy, policy advisers and major institutions and its effect on climate change policy was investigated further in the mid-2000s by political scientist Guy Pearse. He documented a significant revolving door between industry lobbyists for (often transnational) corporations extracting Australia's mineral, soil and water resources; the federal bureaucracy; and ministerial advisers. The agenda-setters shared an economic and cultural ideology, 'a group think' which allowed them to work together inside or outside government.

Previous messages about risk and the need to lower greenhouse gas emissions from fossil fuel use threatened an economic blueprint that, in greenhouse risk analyses, is called 'business as usual' or retaining the status quo. Those messages of risk and proposals for a different energy economy had to be contained and neutralised, a strategy pursued by the revolving door corporate activists and advisers (Pearse 2007).

Radical belief change from social democracy since World War II

It is hard to overestimate the impact of these beliefs on what Australians came to view as normal or reality by 2000. Economic rationalism in Australia, a revival of the English and American free enterprise tradition, had a seismic effect on the values of social democracy characterising Australia for much of the 20th century. Economic regulation in the public interest, (now called 'intervention') along with associations organised for community or environmental benefit to society, would come under increasing attack as being against the economic wisdom and mainstream society.

The mainstream was framed as individual, self-interested working consumers focused on the domestic sphere. The information-fragmenting possibilities of the internet and new media have, arguably, cemented this view of society.

A return to free enterprise, minimal government and libertarian theories are said to re-emerge with the needs of capital (Cahill 2004). The current incarnation was spurred by a slowdown of economic activity in the 1970s following two decades of prosperity after World War II that flourished under Keynesian welfare capitalism. That model of capitalism also promoted mass markets, citizens becoming consumers, and also fed expectations of endless growth while science and technology boomed. John Maynard Keynes's economic blueprint, however, approved of managing consumer demand (upward and downward) and industry regulation to benefit employment and social goals.

Such policies eventually benefitted environmental protection as well, as the first wave of environmental awareness in the 1970s demonstrated. The period after World War II was also characterised by unprecedented support by governments (both capitalist and communist) for science. Technological advances fuelled both the postwar production and consumption boom, and also set up a belief system in the 'techno-fix' for all problems facing society.

Until the 1990s, a social democratic political approach remained strong in Australia in part as an 'accord' between capital and labour. A more inclusive approach to decision-making, including scientists and environmental organisations, was a hallmark of the late 1980s, along with the occurrence of political bi-partisanship for the public interest. The ecologically sustainable development (ESD) workshops were a peak example of this inclusive, mainstream approach.

Climate change communication left with persuading individual consumers

The success of economic rationalist ideas in Australia may have been hastened by the recession of the early 1990s, following similar overhauls in the United States (Reaganomics), and in the United Kingdom under Thatcher in the previous decade. This world view also normalised the goal that societal relations with 'the market' had to be voluntary and by choice. That left climate change communication the challenge of persuading citizens to go beyond a manufactured debate and do the right thing. Paralysis was the more likely outcome, seen again in the recent policy fight over carbon pricing—couched as an impost to the citizen's hip pocket—as if a whole range of responses are not in fact called for.

The individual's prerogative to believe or reject scientific findings flourished in an arena where political leadership had come to mainly express doubt or ignorance and where economic theory emphasised the right to choose for individuals now framed exclusively as consumers or sole traders in a market society. That way of looking at social reality downplayed concepts of community and the public interest and was matched by the downgrading of non-government organisations in the political arena, in line with the thinking of the IPA and other think tanks.

'A disinclination to deal with groups has been reinforced in the major parties by the fashionable theology of public choice theory. This has cast interest groups as selfish and self-serving, and has disputed their representational legitimacy' (Marsh 2005: 222). In this environment, the combined agenda-setting capacity of business lobbyists, politicians and the media encountered little opposition to resetting the climate change narrative during the mid- to late 1990s. Political researchers of this period have documented the 'elites only' decision-making on greenhouse policy of the Howard government after 1996: conferring in serious fashion almost exclusively with corporate leaders of the resource extraction and energy sectors (Hamilton 2006; Pearse 2007).

Belief in 'growth is everything'

In countries dominated by market economics, growth remained the central political objective. US political scientists William Ophuls and Stephen Boyan wrote in 1998: 'Growth is the secular religion of American society, providing a social goal, a basis for political solidarity and a source of individual motivation' (Ophuls & Boyan 1998: 187).

They noted that American politics is a record of a 'more or less amicable squabble over the division of spoils of a growing economy' (Ophuls & Boyan 1998: 187). Even a superficial review of 1990s, and more recent, Australian political economics and culture, exposes the same unexamined, cultural mythology of growth and material consumption underpinning society's wellbeing and prosperity. This thinking governed default responses to the call for climate change action. For example:

US President George W. Bush and Australian Prime Minister John Howard have concluded their bilateral talks in Hanoi, Vietnam, confirming their stance on climate change ... 'We don't believe that Kyoto is the answer' [Mr Howard] said. 'We can have a debate about the severity of the problem, but there is really no debate about the desirability of responding to it, provided we do it in a way that maintains economic growth in our societies and the world.' ('Howard firm on opposition to Kyoto' 2005)

Events since World War II seem to have justified this scientifically flawed belief in endless growth—we see prosperity increased, populations booming, ever more resource extraction worldwide, and human mortality dropping in western societies. There have also been opposing studies and arguments for some time. The 1972 Club of Rome study, *Limits to growth*, applied 'systems dynamics' to economic and ecological trends. It measured trends in resource extraction and the effects on underlying biosphere life-support systems. The assessment concluded (generating much controversy at the time) that 'the limits to growth ... will be reached sometime in the next 100 years. The most probable result will be a sudden and uncontrollable decline in both population and industrial capacity' (Boyden 1987: 217).

The 1990s International Geosphere Biosphere Project (IGBP) studies have shown downward capacity in all natural systems as human activity (the anthropocene) affects the planet. The context is a frequently quoted calculation, attributed to E.O. Wilson, saying that at present rates of resource consumption, if everyone consumed resources at the same rate as Australia or the United States, we would need four additional similar planets to remain sustainable.

Competition policy no help to greenhouse response

Another arm of neo-liberal beliefs in Australia that significantly affected domestic greenhouse policy in the 1990s was the push for National Competition Policy. This was aimed at moving publicly-owned assets and infrastructure into private ownership or at least into a competitive national market. Competition policy, strongly favoured by the Keating government, worked against climate change response through its effect on energy provision and infrastructure, a state responsibility.

With a mandate to corporatise and compete, state-run electricity providers dropped early 1990s commitments to energy efficiency and renewable energy programs and moved into a national competitive arena vying for new customers.

With time, the energy sector became adversarial, and business leaders as well as political leaders became dedicated to 'supply-side' energy management; that is, out to sell more energy for more development and growth.

The simultaneous reframing of the energy efficiency option from 'can do' to 'can't do' was achieved with a 'discourse of inevitability' insisting there were no alternatives to the economic ideas gaining ground (Broomhill 2001). In theory, the 1992 National Greenhouse Response Strategy (NGRS) committed federal and state governments to a range of greenhouse response measures, particularly in the energy sector which, in the mid-1990s, was estimated to contribute 67 per cent of Australia's greenhouse CO_2 emissions and 53.4 per cent of total greenhouse emissions.

The free market direction, however, simultaneously erected financial barriers to efficiency and renewable energy measures. For example, commercial interest rates favoured status quo energy production over innovation and new ventures. Financial barriers also encouraged not putting a price on carbon pollution, which is something economists call an 'externality', i.e. not a core business expense (Walker 1996).

Summing up what he saw happening since the early 1990s, environmental consultant Alan Pears said that whether on energy efficient domestic or commercial buildings, efficient appliances, or transport, Australia during the 1990s and into the mid-2000s experienced an 'almost complete policy failure' in curbing greenhouse gas emissions. He said in an interview:

> We know how to make cuts in every sector, some demonstrably successful. But there are powerful economic groups and narrow theorists and nervous politicians believing that environmental action will hurt the economy. It's been a brilliant PR strategy, and it's left the community confused and disempowered. These beliefs are based on interpretations of crude economic modelling and reinforced by the preconception that you help either the environment *or* the economy.

Pears supports the view that there has been a deliberate political strategy, developed with the resource industries—many of which are multinationals—to discredit the science and scare the electorate with economic modelling on costs. By the second half of the 1990s, energy conservation and fuel substitution had fallen off the policy agenda—in favour of supply-side techno fix proposals such as clean coal, and the nuclear option for a time.

Energy sector deregulation and privatisation with a mandate to compete nationally yielded a prime example of the profound influence of belief and values on Australia's climate change story—with the shift away from a previous state of political/economic bipartisan agreement about the overriding public interest of curbing greenhouse gas pollution.

In terms of the new normal that was created, it's worth repeating that regulation of industrial activity for the greater public good has only recently become a taboo idea. For example, 25 years ago then CSIRO chairman, Neville Wran, told *The Australian Financial Review* that regulation might be needed to achieve emission cuts (McKanna 1988: 4).

This possibility dropped away entirely during Howard's prime ministership as well as from Coalition policies starting in 1996. Australia started a still-unfinished era of the purest expression of economic rationalism and free market capitalism yet seen in this country (Sturgess & Torrens 2009). In terms of climate change response, the public story now rejected 'costs' to markets or regulation to rein in emissions.

Politicians, financial supporters of free market think tanks, and leaders or lobbyists for mining, agriculture, aluminium and electricity generation agreed that there could be no change from the status quo or extra costs to combat climate change. Widespread inefficiencies in the industrial sector, as freely reported in the early 1990s, made the prospect of additional costs more threatening.

Enter the media

Neo-liberal ideas and ways of seeing the world have entered the media mainstream both through the preferences of editors and owners and also via conservative columnists who have been rarely, if ever, connected explicitly to think tanks or what they stand for. For example, while IPA environment director, biologist Dr Jennifer Marohasy penned a column for *The Land* for many years just under her name and shared her sceptical views on climate science with a rural audience without transparent links to her IPA position. Others connected to the IPA have written columns for Murdoch tabloids over the years. In Australia a little of this has gone a long way to setting the public agenda, particularly when these ideas agree with the approach of the government of the day, as we see in the next chapter on the media role.

7. Influences on a changed story and the new normal: media locks in the new narrative

It was the biggest, most powerful spin campaign in Australian media history—the strategy was to delay action on greenhouse gas emissions until 'coal was ready'—with geo-sequestration (burying carbon gases) and tax support.

Alan Tate, ABC environment reporter 1990s

On 23 September 2013 the Australian Broadcasting Corporation (ABC) program *Media Watch* explored a textbook example of why too many Australians and their politicians continue to stumble through a fog of confusion and doubt in regard to climate change. The case under the microscope typified irresponsible journalism.

Media Watch host Paul Barry, with trademark irony, announced: 'Yes it's official at last … those stupid scientists on the Intergovernmental Panel on Climate Change [IPCC] got it wrong', in their latest assessment report. He quoted 2GB breakfast jock Chris Smith from a week earlier saying the IPCC had 'fessed up' that its computers had drastically overestimated rising temperatures. 'That's a relief,' said Barry, and how do we know this? 'Because Chris Smith read it on the front page of last Monday's *Australian* newspaper. When it comes to rubbishing the dangers of man-made global warming the shock jocks certainly know who they can trust.'

But wait. *The Australian*'s story by Environment Editor Graham Lloyd—'We got it wrong on warming says IPCC' was not original either. According to *Media Watch*, Lloyd appeared to have based his story on a News Limited sister publication from the United Kingdom. Said Barry: 'He'd read all about it in the previous day's *Mail on Sunday*,' which had a story headlined 'The great green con'. That tabloid trumpeted about an 'astonishing new admission' and a 'massive cut in the speed of global warming'. Relief indeed.

The only problem was that the error was not the IPCC's but the Murdoch publications' and the shock jocks'. This was pointed out by the University of Melbourne's Professor David Karoly on the same day as *The Australian* story, via a media release through the Australian Science Media Centre. He was joined by John Cook from the Global Change Institute at the University of Queensland noting the dangers of sourcing scientific information from a UK tabloid. The error was due to comparing two sets of figures that were 'apples and oranges', as *The Australian* admitted in its correction a week later. By that time the story had run wild in talkback radio land and in Sydney's *Daily Telegraph* (another News Limited stablemate), despite the expert corrections.

The Daily Telegraph quoted prominent sceptic Bob Carter, formerly a professor at James Cook University, who told the readers the public had been 'relentlessly misinformed'. To make a point, Barry then googled the US space agency NASA to learn that in fact '97% of climate scientists agree' on man-made global warming threats. And, he said, there are any number of other eminent sources *The Daily Telegraph* could have used.

While Carter (a geologist) is out of step with the vast majority of specialist scientists on climate change, the shock jocks regularly interview him to whip up outrage, and he has written a dozen sceptical columns in *The Australian* since 2004 and several in *The Daily Telegraph*, according to *Media Watch*.

Climate change information remains caught up in a manufactured frame of 'scientists don't agree', which matters because the public gets most of its science information from the mass media whether broadcast, in newspapers or on websites. How the media operates in Australia and elsewhere, therefore, has a significant influence on what audiences believe is 'real'. As we can see from this 2013 incident, the ignorant or deliberate sowing of misinformation and uncertainty has not ceased.

Academic studies in the past decade probing public understanding of climate change science and media communication, have tackled the question of what the public is 'getting' from media reports as they are commonly structured, even without deliberate misinformation. The frequent conclusion is public confusion about the causes, effects, risks, and reality of anthropogenic climate change (Corbett & Durfee 2004; Palfreman 2006).

The daily news agenda is an interplay between media and other influential voices—primarily political voices—that guide the dominant narrative for public consumption. Public confusion and apathy are influenced by this dominant narrative and also by public relations strategies applied by politicians and other players using deliberate sceptic language to foster uncertainty.

Mass communication as it has evolved in Australia targets the 'consumer' end of the interest spectrum, and political activity and public discussion have been organised around consumption/economic activity. Rhetoric about costs and hip pocket appeals are potent political tools. Studies have also shown Australian politics is treated by the public as a private choice between political leaders, who are 'consumed' at home via various platforms of media including radio, print, TV and now the internet (Johnson 1987). This is not a profile of an active body politic other than at election time.

Many democracies, including Australia, maintain minimal democratic standards by being accountable to 'the people' through periodic elections. Between elections, however, policy outcomes are determined by elite players in politics, the corporate world and the media.

It's not surprising, therefore, that there is little evidence that public views on climate change drive the agenda, rather the contrary. Australian political science studies since the 1970s have suggested that, in this country, with an elite and top-down political system of governance, the media and politicians together set the daily agenda of what is newsworthy, what is the dominant narrative and what we should accept as real or true. (In the timeframe of the 1990s and early 2000s the internet's main contribution was to allow more voices to add their beliefs and opinions but did not fundamentally change the agenda-setting structure. There has not been a recent body of research showing that has changed significantly).

The frames of commerce and consumption have trained passive consumers

To better understand how the media sets our daily agenda of what is real, a look backwards to influences and analyses from the United States is instructive.

In the late 1960s communications professor Herbert Shiller from the University of California wrote about the connection between mass media and American-style commerce and consumption. The connection is framed as the presence of freedom—in trade, speech, and enterprise. In the war of ideas that has accompanied the resurgence of neo-classical economics since the 1970s, this also came to be framed as freedom *from* government regulation of business on behalf of the public (Shiller 1992).

With global technology, these cultural frames spread, as did their ability to dictate what people perceive as 'reality'. As early as 1951, Canadian communications theorist Marshall McLuhan noted the commercial and propaganda value of the emerging audiovisual media (television broadcasting only gained traction in the 1950s), stating that they provide the viewer with a ready-made image of reality. McLuhan believed that the storytelling devices of mass communication conspired to lull audiences into being passive consumers of culture. He characterised newspapers as the daily 'book' of industrial man, telling thousands of stories to an anonymous audience.

Creating drama with embedded values, hallmark of modern media

Storytelling and personal drama are the mechanisms such that 'even international politics are made a mirror for private passions—love, hate, deceit, ambition, disappointment are the persistent score backing national and international events' (McLuhan 1967: 5). It's not hard to recognise the storytelling features that have come to dominate journalistic practice as we know it: heroes and villains, two sides to every story, and thereby the creation of drama and conflict—with this kind of 'balance' being applied in the 1990s even to scientific stories.

What most journalists think they do for a living, notwithstanding, media companies mass-produce audiences and sell them to advertisers, and therefore they have a major stake in forming attitudes, values and buying behaviour. The public relations and advertising industry has flourished as the mass media turns issues into dramatic stories and, along the way, reinforces the dominant commercial or ideological agenda.

Looking at the image of environmental scientists (including climate scientists) since the late 1980s in this light, it has arguably swung from the more heroic, or at least elite and unquestioned, to that of a fair target of attack. A number of controversial science and society issues during the 1990s, including the science role in mad cow disease, the genetically modified crops debate, and the heated sceptic arguments over climate change, are likely to have been influential in such a shift. A 2010 study by Clive Hamilton exposed where that had led: to the uncivil language of some right-wing columnists towards climate scientists and a barrage of hate email directed against scientists and journalists involved with communicating anthropogenic climate change (Hamilton 2010).

Loss of media diversity equals more capacity to influence

Starting in the 1980s, Australian media became part of a multinational business context with a wave of media mergers resulting in a loss of diversity along with the spread in global business links (Wheelwright & Buckley 1987). Left-leaning economic analyses, like those collected by economist Ted Wheelwright and others into the 1990s, provide a useful historical perspective backed by the documentary record in newspapers and books. They reflect a transition from a more diverse economic and political spectrum of ideas, topics and media publications and programs—to a narrow bandwidth increasingly sounding the same, and dominated by free market conservative ideas.

This singing from the same ideological hymn sheet became more possible as commercial media concentrated in just a few hands (Murdoch, Packer, Stokes as well as the Fairfax family). Australia now has one of the least diverse media sectors in the world (Manne 2005), and it came to be viewed as the norm as the 1990s rolled into the 2000s. The role of News Limited, (the Murdoch press), has been particularly highlighted in this regard on a number of public interest issues, not least its negative position on climate change science and response. Influence is guaranteed with News Limited's near monopoly of print media. 'In terms of circulation, it has almost 70 per cent of the capital city and national newspaper market' (McKnight 2005a: 55–56).

The prominence of conservative commentators in print opinion pages has been matched by a troupe of radio talkback hosts, following John Laws and Alan Jones, who are hostile to climate science. By and large they are framed elsewhere in the media as just a fact of life, rather than as a confusing and misinforming agent in the public discussion whose power has contributed to a disoriented public. Meanwhile public broadcasters, particularly when their funding is in question from conservative governments as happened in the late 1990s and again in 2014, have not taken a leadership position in setting the record straight, but rather have followed the press gallery/political opinion machine in their daily news coverage.

The internet and digital media — losing or gaining common understanding of the world?

As you read this book, social media and digital media platforms have become a dominant factor in public communication. The levelling and democratising effect of the open internet is more evident as a result. With regard to climate change communication, research still needs to be undertaken on the specific effect of the internet on the national discussion. In particular, an open question is the extent to which new communication channels have changed the discussion from an elite and top-down framing exercise to a different paradigm of increasing pressure from the grassroots for change.

It's hard to believe now, but the internet was still in its infancy during much of the 1990s. The early opportunities offered with the entry of blogs, websites and wikis to the public discussion broadened the opportunity for gathering information and retrieving archived information, thereby getting around the limitations of the daily news cycle. It also, had its drawbacks. For every realclimate.org manned by scientists explaining the finer points of climate science and demolishing myths and fabrications, more websites appeared with 'true science' names, manned by sceptics with alternate stories of sunspots and earth cooling, not warming.

How was the public to separate the wheat from the chaff? In the deregulated environment of free market capitalism and the politics to match, where leaders did not lead on this topic and where every citizen was to make up his or her own mind whether they believed in climate change, what did they take from the plethora of opinions that the internet offered? There are signs that audiences are increasingly fragmented as a result, in tandem with media concentration and related internally consistent worldviews.

People increasingly have settled into different information universes, as the adherents of the Murdoch-owned Fox News service in the United States exhibit or, similarly, the more devoted followers of Murdoch tabloids, Murdoch's satellite television channels and related websites, in Australia and the United Kingdom. All of this has made it more than ever a communication story of what people 'hear' and believe.

The extent of influence or competency of 'citizen journalists' afforded by social media applications, such as Facebook and Twitter, in the mid- to current 2000s requires more research. There is evidence of much activity undertaken by young people without waiting for formal engagement with mainstream media, and also by social networks of individuals interested in a particular area, like animal welfare.

One significant trend afforded by new media is the ability to organise global campaigns, like the 350.org disinvestment campaign against coal, involving a call to citizens to remove their funds from banks that support the coal industry. Or the GetUp campaigns to save the Great Barrier Reef from further industrial port development related to coal mining in northern Australia. Such communication avenues relate to news and information but empower citizen action in a way that old media did not.

The traditional model of news gathering and dissemination, however, did not disappear with the advent of the internet, although in 2014 it may be declining. At a 2013 conference in Beijing 'Climate Change Communication: Research and Practice', organised by the Yale Project on Climate Change Communication and the China Center for Climate Change Communication, Andrew Revkin, internet environment reporter for *The New York Times*, spoke on the topic of climate communication and digital media in the west.

His analysis is based on 30 years of experience and concurs with the diagnosis of fragmented messages and audiences:

Until a few years ago the pattern was the same. For a given issue, research was undertaken, papers were written, press releases were prepared, and a related story was composed by a reasonably trained science reporter. When news broke, whether it was the wreck of the Exxon Valdez or the release of a new report by the Intergovernmental Panel on Climate Change, there was a decent chance someone who knew about oil toxicity or the heat-trapping properties of CO_2 would report the story.

That still happens, but less and less. Specialized professional journalists now occupy a shrinking wedge of a fast-growing pie of light-speed media. This reality threatens to erode the already limited public appreciation of science and the state of the planet. I grew up in a world where the media told you 'That's the way it is'. Literally. We all grew up with a common sense of the world. That's not the way it is now. If your concern is climate change, you can go onto the internet and find whatever spin or substance feels like the best fit for your worldview. (Revkin 2013)

He also points out that the world has become swamped with information and 'news', largely thanks to the internet, digital broadcast options and other technological advances, which limits the focus on a single topic like climate change. On the positive side, he uses his own blog *Dot Earth*, which since 2007 has gained several million annual hits, as an example of what can be delivered.

A related phenomenon is that internet technology has enabled information to be increasingly networked and collaborative as 'collective intelligence', linking established news websites like *Time* or the *Guardian* with environmental sites like *grist.org* or political sites like the *Huffington Post* or any number of individual bloggers as seen in Australia. Foundations also support websites; for example the US Center for Media and Democracy, which is focused on correcting the public record. And all of this information is internationally available thanks to the internet.

How mass media habits influence what you 'hear'

What you hear and read, whether in a newspaper or on a blog, can be changed and manipulated by influential political and corporate voices and their public relations advisers, and the task has been made simpler by the traditional news media practice of discovering an issue when politicians make it one. For example, the sudden and much-commented upon arrival of nuclear power in the climate change discourse became an instant issue in 2006 because Prime Minister John Howard and his Cabinet talked about it.

The dominant narrative and agenda of the day are also influenced by the media's internal workings, which are not value-free. News selection in recent decades has often been entwined with ideology related to free market ideas and agendas, according to those who have studied Australian media in the last two decades, particularly the influence of the Murdoch press (Manne 2005; McKnight 2005a).

Professional journalism as a whole has perfected low-risk approaches to newsgathering and called it objective and unbiased journalism. Some common practices have a direct bearing on the communication of science and society stories, and with it the communication of man-made climate change. Australian reporting and editing practices reflect: a reliance on official sources, contrived balance and drama and a lack of context of where the immediate story sits in a longer narrative. The record shows these practices became more pronounced after the mid-1990s in news framing of the climate change story.

The documentary evidence until the mid-1990s shows a more science-focused and contextual coverage of climate change. Thereafter, the predominant media focus in Australia, as in the United States, was the policy/economic debate about climate change response and its 'costs'. Political and economic reporters came to the forefront after 1996. Pulitzer Prize-winning American political journalist Ross Gelbspan, who has written several books on the evolving climate change story in a media context, suggests that the dominance of politics in the evolving climate change story had another internal logic: covering politics has been the perceived elite career path for journalists. Gelbspan sees this as removing scientists and their public interest messages from priority coverage (Gelbspan 2004).

Manufacturing balance

Political scientists Jules Boykoff and Maxwell Boykoff looked at coverage of global warming/climate change during 1988–2002 in nationally read US newspapers. They looked at the effect of 'balancing' an individual media report on this topic. They found, perhaps unsurprisingly, that this adherence to balance—i.e., finding two competing voices, but not necessarily with context—actually biased the coverage of anthropogenic climate change and issues related to response, and led to uncertainty (Boykoff & Boykoff 2004).

There are historical precedents for manufactured uncertainty and argument infecting controversial social issues from slavery to cigarette smoking. With this winning strategy, uncertainty exploited by sceptics in contact with media editors has met with gratifying success in recent decades (Oreskes & Conway 2012). The success has been to establish a need for scientific balance on climate change—requiring that dissenters from the mainstream science be given equal time, even if they only represent themselves or one per cent of a scientific field.

Frequently the sceptical voices have not been scientists—as in a September 2000 article in *The Australian Financial Review* by a 'former senior public servant' who quotes a small handful of armchair Australian sceptics, such as Tasmanian John Daly, and tells his readers that 'scientifically the greenhouse scare is largely over' (Scott 2000: 34).

Context avoided or used strategically

Context is often avoided in news reporting and thus is another confusing influence on climate change stories. This can happen to forestall a charge of bias against the reporter who might add the context. Applying context is also more time consuming, as it requires research and/or experience. Instead, the standard, 'objective' and balanced approach is reflected as a 'he said', 'she said' array of facts and opinions—possibly assembled in a number of consecutive news stories, but without background as to where this information fits in the ongoing evidence or science discovery process.

Context is often also missing for people quoted or interviewed—do they speak for a peer-reviewed research summary in the case of scientists, are they a relevant expert or are they a sceptic representing a non-peer reviewed minority opinion and come from a non-expert field? A related omission is an interviewee's affiliation or research background. Rarely does a news report include either a sceptic's or a mainstream scientist's credentials or a relevant link to a particular interest group or think tank. These omissions encourage the common public notion that scientists are interchangeable. 'Scientists say …' is a standard introduction.

Experimental work has shown that news consumption without context does not lead to better public understanding. In testing a sample audience with various treatments of a global warming story, a research team found controversy added to readers' confusion—while context made people feel more certain that they understood global warming and that it was real but complex (Corbett & Durfee 2004). Only rarely is context used in modern news bulletins, as the journalism profession requires, for better understanding.

Fragmented news disabling

The fragmented format of modern news presentation of long-running stories disables good understanding of public issues. Thirty-second sound bites, and the belief that the public has no concentration span, have not helped.

Professional journalism tends to pummel people with facts, but rarely pummels people with a nuanced appreciation of what the facts might mean. This helps explain the numerous studies that show that sustained consumption of the news on a particular subject often does not lead to a better understanding of the subject and sometimes leads to more confusion. (Nichols & McChesney 2005: 19)

Instead of simultaneous analysis and context, drama and manufactured 'balance' have become the staple formula for news stories, and this has crept into reporting of some scientific and environmental issues, creating controversy. It's not a long step from creating balance to relying on duelling opinion. This appears to have happened with the two media organisations that I assessed.

They both changed their coverage incrementally, offering science and economic coverage of the climate change story that continued the risk and human agency understanding from the late 1980s story into the mid-1990s, as governments changed. They shed their editorial certainty on the matter by the end of the decade, however, to differing extents.

Thirty *Sydney Morning Herald* reports sampled from the second half of 1995 and the first half of 1996 on the topic of climate change almost all focused on the science and risk messages as well as on international negotiations or the Australian economy's dependence on coal exports. For example, headlines included: 'Malaria spread linked to climate change', 'Climate change a fact: experts'—a report syndicated from *The New York Times* about the 1995 IPCC assessment; and a January 1996 report with the headline 'Plummeting penguin numbers a crisis on Macquarie Island' explained wildlife losses as Antarctic waters warmed. There were no sceptic opinion pieces in these samples.

'Australian ploy fails to slow greenhouse action' by *Herald* technology writer Gavin Gilchrist, writing at the close of the Labor government under Paul Keating, is a noteworthy example of reporting that persisted to 1996. It gives a fair sense of the direct and unambiguous approach of an earlier phase of climate change journalism, which does include scientific context with a political report. Gilchrist wrote:

Australia has sought to weaken international efforts to tackle the greenhouse effect by trying to undermine a landmark scientific report that calls for immediate action to ward off global climate change.

It is the third time this year the Federal Government has tried to delay international action on the greenhouse effect: in March, a botched diplomatic strategy at the Berlin climate convention was not adopted, and in August it emerged that the same diplomatic strategy was being pursued using an economic study partly funded by the coal industry.

(three paragraphs later)

> For the first time, the world's governments will be advised that the risk from climate change is so great that immediate action is warranted beyond measures which make economic sense, such as improving the efficiency of energy use by industry. (Gilchrist 1995a: 1)

In the same year, Gilchrist also wrote about a CSIRO report with the headline 'Greenhouse effect will cause havoc in NSW, study claims'. Increased risk of severe thunderstorms and torrential rains are a prominent theme of this report, again providing evidence that likely impacts were understood and reported early. Amongst the CSIRO findings Gilchrist wrote:

> Sydney will suffer twice as many days of extreme heat, four times as many severe storms and far worse flooding from huge increases in torrential rain, according to the latest predictions of how NSW will fare under the greenhouse effect. (Gilchrist 1995b: 5)

Herald reporter Bob Beale examined the planning process for coalmine development in New South Wales and offered graphic statistics on the impact of Australia's coal focus. He wrote that, although 'it would take 420 million new trees to soak up the estimated 281 million tonnes of greenhouse gases produced by the Hunter Valley's proposed Bengalla coalmine, according to calculations by a Federal Government bureau', mine-lifetime greenhouse emissions were not being assessed as new mines were opened (Beale 1996: 9). In this case the information source was a government report that put coal mine development in a greenhouse gas context. In the the mid-1990s *Herald* sources generally were still scientists or politicians.

From 1996 and the government change to the conservative Liberal and National parties (the Coalition) for 11 years thereafter, the evidence indicates that climate change became increasingly framed as a political/economic story with a strong component of Australia blocking climate action internationally. Again we can turn to Gilchrist who reported in July 1996 that:

> The Howard Government today steps up its diplomatic offensive opposing international efforts to protect the world's climate at the historic meeting of the Climate Change Convention in Geneva.
>
> Australia, with its pro-industry stance, is set to be seen as a rebel nation out of step with mounting global concern about the threat of climate change from the greenhouse effect. (Gilchrist 1996b)

Unlike many later political stories, Gilchrist does not ignore the scientific context of risk inherent in heating up the planet with greenhouse gas emissions. Let's remember that science and environmental reporters up to this time did not treat human agency in the warming as a debate, but rather as the fundamental cause. He cites the 1995 IPCC report, plus a background report on Australia's preferred economic direction of backing existing fossil fuel-based energy producers and users.

The article also flags the return to a traditional 'economy *versus* environment' policy framework that characterised the Howard government's pro-industry stance on downplaying greenhouse gas emissions. Gilchrist describes the revolving door of like-minded executives cycling between bureaucracy and industry and setting the political agenda, a story retold by Guy Pearse a decade later in his 2007 expose of the so-called 'Greenhouse Mafia' of lobbyists and policymakers.[1]

The *Herald* continued to run well-informed and in-context science stories during the later 1990s, but a framing shift became apparent during this time. Along with the change of government in 1996, there began to be relatively more political/economic coverage in the lead-up to the Kyoto climate change summit. This can be seen starting from 7 June 1996 and during the following six months where a *Herald* editorial and five of six articles focused on international negotiations.

In comparison one finds that amongst 30 articles sampled from 1995–1996 for *The Australian Financial Review*, only four are focused on the science. One is on a technical issue and is an opinion piece (i.e., not written by a staff or freelance journalist) and three are opinion pieces by sceptics—by US scientist Michael Patrick, National Party politician John Stone and former Labor minister Peter Walsh (later to become a prominent member of the ultra sceptical Lavoisier Group).

The same sceptical approach to the science was not the case in sampling from 1987 to 1992, when the *Financial Review* ran a mix of straightforward science reporting (defined as quoting mainstream climate scientists), and political/economic stories, some of which were candid about Australian industry's inefficiencies.

By 1995, however, the economic concerns of energy producers and big electricity users, like aluminium producers, predominated in the coverage, along with the focus on international negotiations. Unlike Gilchrist's stories for *The Sydney Morning Herald*, the context of scientific assessments was no longer reported in this publication.

Besides politicians and sceptics, sources in these later *Financial Review* articles frequently include industry spokespeople urging the government to heed their

---

1   Gavin Gilchrist quit reporting in the late 1990s to promote sustainable energy, which he wrote about in an opinion piece in 2001. Several other reporters interviewed for this story had moved on by 2001 to private careers in sustainable energy promotion or state government. This is another influential structural feature of news media—the loss of experienced and well-informed personnel.

concerns or agreeing with government about what is in Australia's 'national interest'. Conservation group spokespeople are posed in opposition. By this time green groups rather than scientists were more often making the case that climate change is a problem.

Many *Financial Review* headlines in 1995 relate to fears of a carbon tax by mining and energy producers and providers, or indeed, of any tax or international regulation to limit emissions. Here is evidence that even before the change of government in 1996, industry and government had already been reframing Australia's position from the early ethical, response-focused and internationally cooperative stance to an economic self-interest stance that ignored the risk messages.

Headlines included: 'Business in last ditch bid to bar carbon tax', 'Australia takes strong line against greenhouse rules'; 'Business lines up to fight controls' with a 'party line' of quotes from industry spokespersons. Similar framing can be found in 'Macquarie fears a greenhouse handicap' (Callick 1996b) where the rhetoric of the investment sector is quoted and this is later countered by a Greenpeace spokesperson. Callick wrote:

> Policy options on combatting climate change that are still before the Federal Government 'could destroy the competitive advantage of Australian mineral processing companies', according to Macquarie Equities Ltd …

> Australian energy and commodity producers would come under increasing pressure to conform to the policy stance of Europe and the US as negotiations proceeded. … The Europeans' position was driven by trade competitiveness objectives, the Americans' by the presidential election.

The construction of a report featuring either an industry point of view or a federal government point of view or both countered by an environmental group later in the story is common amongst these *Financial Review* stories. This story formula reinforces the frame of the 'mainstream', which is represented by industry and government looking after the 'national interest', versus 'special interest', the environmental group opposing business and jobs.

The value frames are familiar and still being used today: the rhetoric about Australia's competitive advantage, that if action is taken industry will go offshore. This was often coupled with the idea that outsiders were working against us ('us and them'). European or United Nations attempts to progress emission reductions have been framed as self-serving and not in Australia's interests. In this way political and economic reporting was establishing the dominant narratives.

News is what powerful people say

The economic and political articles on climate change of the 1990s illustrate a common contemporary media practice: stories are framed as authoritative and 'objective' when they report on what people in power say and do. But those people in power change and with them often the narrative. Reporting political utterances is presented as removing bias from story selection, and it makes newsgathering less expensive. News companies set up reporters near powerful people; for example, in parliamentary press galleries. While this may reduce journalism to networking and scribing or opinion pieces, it is safe.

In Australia the federal parliamentary press gallery dominates the daily news coverage. On any given day political back and forth is the majority of what is relayed by the national broadcasters, the ABC and the Special Broadcasting Service (SBS), which has a multicultural focus, as well as the major dailies. Understanding this political reporter–politician nexus makes the agenda-setting role of governments *with* the media more transparent. It also sheds light on why political leadership becomes so important in the will to action on a controversial science story.

In this media environment politicians are primary framers of the climate change topic. By the 2000s, when no politician was talking about climate change, editors were liable to assume there was no story, according to the accounts of two Fairfax journalists (one working at *The Sydney Morning Herald* and the other at *The Australian Financial Review*) who wrote separately about their on-the-job observations in the professional journal *The Walkley Magazine* (Frew 2006; Macken 2006).

The internal structure of newspapers also influences whether a story 'gets up'. This is related to the power of the editor and often is subject to the values of the editor. Editors are appointed because their values are coherent with the dominant ideological culture of the media group ownership and board. This coherence often extends to national policy as we see with the communication of global warming and climate change.

During the later 1990s and since, with the private media concentrated in only a few hands, and the often timid public media sticking close to the narrative set by the loudest voices from the press gallery, the public was getting a unified message from its news media: the main game was no longer avoiding catastrophic climate change by reducing emissions. That was uncertain and too costly. Now the story was all about Australia's (narrow) economic drivers and international 'comparative advantages' relating to resource extraction.

This refocus from a mainstream science story to a political focus of 'can't do', because of costs and national interest, shows us how an ideological framework is imposed on a society through communication to change its sense of reality.

Under this new 'normal', the status quo fossil fuel industries—coal, oil, natural gas and their derivatives for industry—had assumed unquestioning 'must have' status. There was no more criticism of inefficient energy use by commerce and industry such as was seen up to the early 1990s. This perspective also overshadowed the earlier interest in alternative energy production. Those were now labelled an ineffective sideshow.

Defending fossil fuel exports and domestic 'cheap' coal power, led Australia to portray itself as exceptional in international climate negotiations. The 1997 Kyoto Protocol was the means for establishing proposed emission boundaries. But Australia positioned itself by the early 2000s as a principle opponent of *ratifying* the protocol with its mandatory targets for restricting emissions.

The inherent difficulty for conscientious reporters in that framework was noted by one environment reporter recounting the experience at *The Sydney Morning Herald* by the mid-2000s. She found that after the first decade of the federal Howard government, political correspondents and editors were uninterested in the topic except as an international battle (Frew 2006: 18).[2]

A similar observation comes from former ABC television environment reporter Alan Tate, who saw firsthand the changes that occurred in the 1990s: the influential (to editors) Canberra press gallery took its climate change information from the government. A former journalist from *The Age* metropolitan newspaper in Melbourne, who took the subject seriously and responsibly in the mid-1990s, told me that when she suggested a climate change story her editor responded: 'haven't we fixed that?' She also said she was labelled a 'greenie'. Another reporter who covered climate change for the *Herald* in the later 1990s, Murray Hogarth, said: 'We were a lonely bunch in the 1990s—I knew of no editor who was committed to telling the story.'

---

2   The federal parliamentary press gallery in Australia generally ignored climate change in the late 1990s, as did the politicians, apart from the political battle over the Kyoto Protocol. This continued well into the 2000s, until recently (Sanderson 2006). How it might be different was shown with the avowed turnaround of News Limited owner Rupert Murdoch on the topic in 2007 (reported nationally and internationally in May 2007; e.g., Griscom Little 2007b). We suddenly find *The Australian's* national affairs reporter cutting through the government rhetoric and giving a cogent account of the real economics attached to mitigation, although the headline 'Green row will be decided on economic fear' still makes it sound like a green sectoral issue (Steketee 2007). This understanding, however, did not last in the News Limited daily print media, which soon thereafter continued with a critical and sceptical public stance.

Opinion and uncertainty: media hallmarks by 2001

By the time of the 2000–2001 IPCC reports, 30 relevant articles sampled from both *The Sydney Morning Herald* and *The Australian Financial Review* for those two years document the shift to opinion pieces 'balancing' the science along with continued political coverage from the perspective of the new narrative. Content continued to change towards uncertainty.

The *Financial Review* was regularly calling global warming/climate change 'a debate' in the 2000–2001 stories and continued to quote sceptics as the science context. The common trend by this time to dismiss alternative energy solutions as marginal and non-mainstream can be seen in *Financial Review* reporter Nick Hordern's (2000) piece. He describes renewable energy as a manifestation of 'green politics' and 'subjectivity' that 'few energy analysts' agree with.

I compared 30 *Herald* articles from 1988–1989 with 30 articles from 2000–2001. I found that the number of opinion pieces had gone up tenfold by 2001 from a level close to zero a decade earlier. Journalists' reports about climate science in the period 2000–2001 were at times placed on or near the opinion pages, where sceptical tracts also appeared, if not always on the same day. This emphasised debate, opinion, and uncertainty about who to believe. The samples also suggested that by the end of 2001 the newspaper was quoting green groups and non-government organisations (NGOs) three times as often as in the early days.

The use of scientists and experts as sources declined by about 20 per cent from the early comparison period. This is consistent with the impression that green groups were quoted more often in an adversarial role to the government's position. Named politicians and industry spokespeople as sources had, however, also gone down based on this sample, with industry comment showing up as statistically negligible in the sampling periods. This was not so for *The Australian Financial Review*, in which the industry point of view increasingly set the agenda, as articles from 1996 on show.[3]

---

3 Sceptical treatment was not consistent in *The Australian Financial Review*, despite the general trend in that direction from the mid-1990s, possibly adding to reader confusion. At the time of the 2001 IPCC report, several stories appeared, including about the insurance industry's concerns, that were framed as certain about climate change and its connection to the fossil fuel-based economy. For example *Fossil Fools* (Huck & Macken 2001).

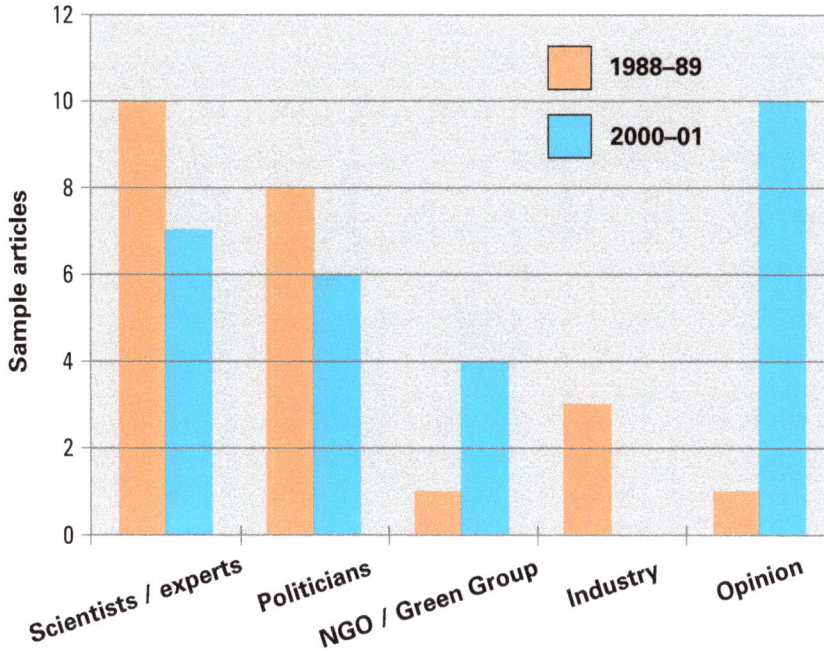

Comparison of sources for *Sydney Morning Herald* articles 1988–1989 and 2000–2001.

Source: author's compilation.

Sydney Morning Herald science reporter Deborah Smith's story on the draft 2001 IPCC assessment on 13 November 2000 was notable as a science update placed in the paper next to the op-ed features section, where sceptic pieces also were placed. Furthermore, Smith's article was introduced with the words 'The relations column will return next week'. The subject matter was not opinion but a straight report on the 2001 IPCC assessment. Smith wrote:

> (The IPCC) tone has toughened considerably since (1995), based on new studies. The latest draft report by the 3,000 scientists who make up the Intergovernmental Panel on Climate Change, IPCC, to be finalised early next year, warns that mankind has 'contributed substantially to observed warming over the last 50 years'. (Smith 2000: 13)

Smith also interviewed Graeme Pearman then chief of the CSIRO Division of Atmospheric Research who stressed that the underpinning science was solid and response action should not be delayed on behalf of 'a few remaining greenhouse sceptics'. In this November 2000 issue, the page one story focused on environmentalist dissatisfaction with the Australian position at the Hague climate change conference that month. The headline 'Greens flex their muscles at "last chance" climate summit', underscored the government versus the greens frame that the narrative had assumed.

Smith's piece was preceded on the same pages on 19 October by a piece called 'Hot news, the greenhouse effect is not so bad after all' by Larry Mounser, who was credited as 'a freelance writer, a physics teacher and runs a course'. Mounser also contributed again later in the year.

In the October piece he argued that, since the climate in recent geological time is a series of ice ages with brief interglacials: 'The onset of an ice age could take just 70 years. Being able to avert it by burning fossil fuels, purposely creating a "greenhouse effect", could be one of the luckiest flukes in human history. Yet, strangely, it's the warming of the planet that we fear'. He wrote that documented natural temperature variations with the Arctic losing ice have caused no ecosystem harm and furthermore 'there's also no hard proof that CO_2 is causing the warming anyway' (Mounser 2000: 12).

On 6 December, after Smith's IPCC piece, Mounser appeared again writing about 'Cracks in the greenhouse ...' claiming hundreds of studies don't support the IPCC, and offering some plausible (to the layperson) alternative perspectives. He used rhetoric like 'the high priests' and the 'white coated posse' when referring to mainstream climate scientists, and this name-calling would be a growing trend.

The *Herald* samples for the next year, 2001, presented fewer opinion pieces. Instead the paper ran a stream of political stories about why the United States, supported by Australia, did not want to ratify the Kyoto Protocol. There were also some informative stories linking weather outcomes to climate change. In January 2001 the IPCC assessment, now officially released, received page one treatment, with the risk message being emphasised in the headline: 'Six degrees hotter, global climate alarm bells ring louder'.

What was missing was the context that this alarm was raised more than a decade ago. The 'news' tag applied to successive IPCC reports implying that the risk to society and human agency had just been discovered or made more certain, was in fact historically inaccurate and served to dull urgency about a problem that had been known in detail for 20 years and longer.

Leading audiences to assume that things were only beginning to become scientifically clear also made it harder for them to 'connect the dots' with extreme events on the ground. Instead in 2001, the focus stayed on debate: international response negotiations were framed by both reviewed newspapers as a battle between environmentalists and the government backed by industry. A *Financial Review* article in November 2001, 'Conservationists fail to expel Australian team', for example, reported on non-government organisation complaints about the Australian negotiating team and its role in Kyoto Protocol negotiations. In that article, the government's role is defended by the Australian Aluminium Council's representative and public relations specialist John Hannagan.

Taken together, the 2000 and 2001 articles suggest that the discussion might be dismissed by audiences as uncertain and debatable or opinion, or as a special interest issue of little concern to the mainstream.

Commercial pressures and ignoring the dots

Commercial considerations always help shape how news is presented and what the important stories are for the ever-more concentrated mass media owned by a few giant corporations (Bagdikian 2004). Commercial pressure regarding the climate change story is therefore nothing new: Gelbspan reported that more than a decade ago in October 1999, he had a conversation with a top editor of a major US TV network asking why the dots were not connected between increased coverage of weather disasters and climate change. The editor said on the one occasion where they tried it, a barrage of complaints was aimed at the top network executives from the industry-funded Global Climate Coalition.

The fossil fuel industry argument then, and now, amplified by sympathetic politicians, is that any one event cannot be linked to human-induced climate change; even mentioning that scientists linked a pattern of violent weather with climate change has been deemed offensive. The editor said the network was intimidated. 'The threat was implicit: if the network persisted, it ran the risk of losing a lot of lucrative oil and auto advertising dollars' (Gelbspan 2004: 80).

In Australia, ABC reporter Alan Tate said that his editors were also deluged with complaints from the resource extraction industry whenever he covered climate change during the 1990s. While not a commercial threat, it might be considered a political threat to the public broadcaster.

It is noteworthy that with recent extreme weather events in Australia, such as the cyclone that destroyed Innisfail in 2006, the 'Black Saturday' extreme bushfires in Victoria in 2009, and the extreme flooding and cyclones of 2011, fire devastation in Tasmania and New South Wales in 2012 and 2013, a similar lack of 'connecting the weather dots' remains evident from politicians and media.

Enormous spin campaign succeeds with reporters

Structurally, the way a newspaper or TV news service is organised is in 'rounds', and many science and society issues like water, wildlife management and climate change will be covered across different rounds—political and economic, as well as science and environment. This affects the quality of reporting: often resting on the judgement of non-science reporters and their willingness to accept opinions or propaganda as fact without a critical knowledge base.

Veteran American journalist Bill Moyers, now one of the corporate media's sternest critics, described the dominant political journalism culture in the United States—and it can as easily apply to Australia:

> 'Instead of acting as filters for readers and viewers, sifting the truth from the propaganda, reporters and anchors attentively transcribe both sides of the spin—invariably failing to provide context, background, or any sense of which claims hold up and which are misleading.' (quoted in Nichols & McChesney 2005: 25)

'Spin' was also the word used by Tate when he said high-energy users and production industries—aluminium, coal, electricity, and later fertiliser and cement—were actively setting the climate response agenda along with federal officials. Reporters heard and amplified a steady narrative of 'go slow' on climate action (also documented by Pearse 2005 and Hamilton 2001). In Tate's view, 'It was the biggest most powerful spin campaign in Australian media history'. He understood the strategy was to delay action on greenhouse gas emissions until 'coal was ready'—with geo-sequestration (burying carbon gases) and tax support. He told me that what he saw of the communication tactics was:

> First sow seeds of doubt about the science — make it a nonsense. Say let's not be part of the Kyoto Protocol — it's too little anyway. Then say OK we've got a techno fix, geo-sequestration and nuclear. Ignore energy efficiency and renewables, why bother, those are green issues, it's all marginal. The Oz main game is coal and cheap energy.

The historical record of this period, found in published documents and books already mentioned, largely agrees with Tate's assessment. What was accomplished during the 1990s by corporate interests, politicians and media was the reframe of the story by denying, downplaying or confusing the risks of climate change and the cementing of a 'business as usual' narrative. The resource industry and lobby groups from aluminium to forestry played a strong role behind the scenes. Political and economic reporters ate it up. Former mass media journalist Wayne Sanderson wrote on the political website *Crikey* in 2006 about the federal press gallery during the Howard government from 1996 on:

In attempting to dictate the terms of the response to climate change, John Howard is the doctor who denied the disease, but now wants to prescribe the cure. And the press gallery shows every sign of letting him get away with it. In fairness, the gallery may be doing the best they can, given they are intellectually retarded on this subject, having shown little interest in it over the years. Search the archives, in vain, for a serious piece by a serious 'insider' on what has been a monumental failure of national public policy.

As a pack, the gallery has allowed the climate change debate to be framed by the government—first it wasn't happening; then it was happening, but there wasn't much Australia could do; now it is serious and nuclear energy will fix it. At each point, the stance has been either totally wrong, or at least questionable, but the fourth estate has been missing in action. (Sanderson 2006)

Where was the public broadcaster and the public interest?

With the parliamentary press gallery missing in action or at least missing the point, the next question might be to enquire more closely why Australia's iconic public broadcaster retreated from informing the public about climate change during the course of the 1990s. Veteran ABC reporter Allan Ashbolt gives one answer, contending that the ABC's main function is 'to legitimise and stabilise the culture and ideology of the present socio-economic system' (Ashbolt 1987: 14–15) and that the 'ABC passively accepts the ideological values passed on by outside institutions'.

In other words, even without the commercial imperative, did the public broadcaster just reflect prevailing government ideology and the narrative agenda set by politicians and the dominant parliamentary press gallery journalists in any given period? The evidence suggest that might be the case, and there was a noticeable switch in the climate change coverage by the end of the 1990s, after an earlier period in the news and science departments which had covered the risk and responsibility story fully.

Looking back, former ABC broadcast producer Richard Smith recalled that he and journalist Geoff Burchfield produced a half-hour thematic special on climate change for the science program *Quantum* in 1988 and that the special came about because the science reporters and producers decided to 'force the issue' and sold the program to management. Asked what triggered their interest, he said the science of climate change was 'common knowledge' and emerging as 'a serious scientific issue' at the time.

Smith also noted that conflating the hole in the ozone layer and climate change science was a misperception at the time that the program tackled. A specific trigger for the *Quantum* special was the scientific work coming out of the CSIRO by Pearman and others around the time of the first 'Greenhouse' conference in 1987.

A four-season series followed with *A question of survival*. By the early 1990s, however, management sentiment had changed within the science unit. Smith was not sure where the pressure to drop the program came from but the production team was told the audience mood had shifted and people were not interested in an environmental science series anymore. At the same time interest in *Quantum* science coverage had not changed, he told me and, 'it didn't seem to me from my interpretation of the figures that there had been a serious erosion of audience interest in environmental science matters'. Management, however, became more interested in the growing information technology revolution.

Meanwhile, ABC news division continued to support Tate's reporting well into the 1990s, he said. He left in 1998 when he felt that deep uncertainties about climate change had settled into the editorial policy. Other contemporary observers and insiders of the ABC testify that in the late 1990s and early 2000s, self-censorship and timidity marked the public broadcaster *vis a vis* federal government narratives on controversial issues (Manne 2005).

The fact that we still hope for more consistent and incisive journalistic standards from public media may thus appear optimistic, but sometimes that hope is still realised in the remnants of investigative journalism found in ABC *Four Corners*, *Media Watch* and some Radio National documentaries.

Culture wars, agenda-setting, and owning most of the media

'Culture wars' is a label for attempts to influence the dominant ideas and values driving a society. Media moguls can have that influence and Rupert Murdoch is one with a long reputation for culture warring and agenda-setting with a deregulation, free enterprise and 'freedom' mission. His News Limited (and News Corp internationally) has shown its ability to help make and break governments not only in the United Kingdom, but also in Australia, as has been examined in a shower of recent books and documentaries.

Parent company News Corp has an international reputation for inflammatory and opinionated publication with significant right-wing influence in the United States with the *Fox news* network. It also exerts ideological influence through the *New York Post* and *The Wall Street Journal*. In the United Kingdom, News Corp enjoyed unparalleled political influence until undone by the recent phone-hacking scandal enveloping its tabloid press, which also dampened monopoly plans for its satellite Sky TV network.

News Limited Australian mastheads include the national daily *The Australian*, as well as *The Daily Telegraph* (Sydney), *The Herald Sun* (Melbourne), *The Courier Mail* (Brisbane) and *The Advertiser* (Adelaide). Sky News is here as well and making inroads on the mobile communication front.

This dominant newspaper position in recent decades, in combination with political attacks on the ABC (and the broadcaster's timidity in response), plus an aggressive right-wing commentariat assembled on News Limited pages, has had profound impacts on Australian political culture. Not missing has been significant influence on public understanding and discussion of anthropogenic climate change (Manne 2005, 2011).

The Australian and the tabloids have maintained a consistent stance since at least the mid-1990s of uncertainty and doubt regarding climate change science and any attempts at regulated response that would affect commercial activity (Manne 2011). The world view coincides with the increasing hegemony of market fundamentalist thought in Western English-speaking democracies where New Corp operates.

A newspaper can exert influence and reflect its perspective in editorials, opinion pieces, headlines and story slant. In the case of *The Australian* and its 'culture war' on this topic and others, journalist and media researcher David McKnight writes that, editorially, News Limited exhibits a sympathetic value relationship with right wing think tanks and federal politicians from both major parties:

> This orthodoxy is one which holds to certain doctrinaire ideals about economic management, national identity, foreign affairs, public schools, climate change and many other issues ...

> It is an orthodoxy which is shared by a number of senior journalists at News Ltd and by many of their editorial writers, columnists and contributors. It is an intellectual universe in which a network of conservative think tanks, academics and writers of the right have a symbiotic relationship with the newspapers of News Ltd ... and most significantly the coalition government under John Howard. (McKnight 2005a: 54)

Based on sympathy of ideas, it is not surprising that Murdoch backed, with his newspapers, the Howard Coalition bid for government in 1995/1996, and Tony Abbott in 2013, while in 2007 Murdoch reportedly thought he would have major influence with Kevin Rudd and backed him at that time.

A near monopoly of news outlets in major metropolitan centres can quickly spread ideas it favours. An analysis of News Limited publications exposed their positive coverage of geologist Ian Plimer's 2009 contrarian book on climate science (McKewon 2009). The researcher thought this might have had a significant effect on the political discussion of an emission trading scheme at the time. She also documented a trend to release sceptical science books at key political time points.

This has been an ongoing tactic of free market think tanks that question climate change, particularly the Institute of Public Affairs (IPA) and the Lavoisier Group, with their thoughts amplified by sympathetic media outlets. Four such books were reportedly published in 2009, including Plimer's. (The public broadcaster, with its national coverage, also gave Plimer's book a sympathetic airing as a 'balancing' voice).

McKnight gives examples of the undisguised ideological bent of some News Limited editors and writers over the years (in a profession that cloaks itself in supposed neutrality and impartiality). He writes of editor Leslie Hollings who had open ties to the IPA in the 1980s: 'For a decade Hollings was a key figure in fashioning the ideological stance of *The Australian* and News Ltd' (McKnight 2005a: 61), including championing economic rationalist goals like deregulation.

In this way 'culture wars' of ideas are deployed on the value front, with editorial policy slanted for business and political objectives on an issue like climate change. Who first marks the agenda—whether media or policymakers—may be difficult to unravel: in the climate change case it may well have fallen into place through personalities and ideological agreement.

But in 2005 at least one veteran newsman and long-time ABC professional, Quentin Dempster, concluded that the lack of diversity and the concentration of media power in Australia was a sign of media corporations' power *over* the politicians, affording a handful of media owners strangleholds over the nation's sources of mainstream information. He wrote: '... we must remind ourselves that Murdoch (News Limited) and Kerry Packer (Consolidated Press and TV channels) are not called 'the gatekeepers' for nothing. They have had a testicular hold on our prime ministers from Fraser to Howard' (Dempster 2005: 113).

To put this in further context, influence lies not just with global media companies such as News Corp. Transnational corporations in general exert major influence in Australia and did on Australian climate policy during the 1990s—aluminium smelting, coal, and metals corporations topping the list. These are backed by US-developed public relations techniques and the agenda-setting role of domestic and international free market think tanks since the 1970s.

Regarding other Australian print media, media critic Guy Rundle (Rundle 2005) painted a picture of journalistic 'decline' at the Fairfax corporation, which owns *The Sydney Morning Herald* and *The Australian Financial Review*. This began under the influence of a conservative board led by free market competition policy advocate Fred Hilmer and prominent conservative businessman (sports, mining, casinos) Ron Walker in the mid-1990s.

Rundle characterised the decline as a shift from being a publisher reflecting a pluralistic society to one more closely aligned with the economic rationalist world view that overtook Australian culture and society. That may help explain the change in treatment of climate change stories and the shift to uncertainty through opinion and 'balancing' of stories in the later 1990s in the *Herald*, and the shift to a partisan defence of resource industries' interests in the *Financial Review*.

Since then the Fairfax empire has suffered major economic woes and contraction, having fewer resources than News Limited to weather the worldwide decline of print media related to changes in advertising income. In its slimmed down state, however, Fairfax has in recent times produced significant climate change science and policy coverage that is more akin to the early days.

Language that supports culture wars

A similarity to classic propaganda techniques is evident with the 'us and them' framing that has taken place in the course of redefining the climate change story. Divisive language denigrating climate scientists and environmentalists has been part of the armoury of this culture war, while the general rightward shift has seen regulation of business for public health or environmental concerns painted as a drift to socialism and to the 'left' influenced by outsiders, without raising many eyebrows.

The divisive 'us and them' frame is regularly applied by columnists and particularly radio talk show hosts to decry climate change science and those who believe climate change is an urgent risk to society. Take for example a 2009 piece by columnist Christopher Pearson, a former speechwriter for Howard as prime minister.

In launching Plimer's book in *The Australian* under the headline 'Sceptic spells doom for alarmist religion', Pearson accuses climate scientists and environmentalists of 'religious' fanaticism and calls the IPCC findings pseudo-science led by 'eco-fundamentalists' who hate the modern world and subscribe to 'anti-human totalitarianism' (Pearson 2009). 'Us and them' rhetoric is not confined to any one sector or ideological perspective in society but is relevant to this story because of its contributing role in re-establishing neo-conservative and traditional values from 1996 on.

Story metaphors of home, hearth and national interest

Getting the story wrong in the news media has been made easier by the journalistic convention of reporting issues as dramatic personal narratives, framing global issues such as anthropogenic climate change in metaphors of personal loss or gain and one-on-one contested argument. Thus the dominant narrative by federal politicians and the media in the late 1990s painted action on climate change not as risk management for the whole society, but as a threat to jobs and businesses, and an attack by 'them' (in Europe or the United Nations) on 'our national interest' (read family).

In this storyline, market capitalism is synonymous with political democracy and 'freedom', and there is a natural order in the type of economy Australia operates—i.e., the emphasis on export of natural resources. Freedom evokes a metaphorical pathway that signals choice and lack of regulation and 'national interest'; that is every family's interest, comes through wealth from mineral extraction—coal being most relevant here. These value metaphors gloss over the reality that co-driving Australia's resource extraction policies are multinational corporations with their own interests, both on the ground and in the media.

How this coded value language is applied in the media/political culture wars is further illustrated in a 2003 book by David Flint, *The Twilight of the Elites*. Flint enjoyed a power position in regard to the Australian consumer and the Australian media as former chair of the Australian Competition and Consumer Commission (ACCC) and of the Australian Press Council during the 1990s. His book candidly sketches and approves of a campaign by the right against the left wing of Australian politics.

Australian sovereignty and prosperity are dominant themes in this world view, as is an attack on 'Australia's media and legal elites' represented by those who disagree. Labelling opponents elites (as opposed to the rest of us) has been a common rhetorical tactic. He writes: 'A significant feature of the elite agenda involves the surrender of part of our sovereignty to international organisations' (Flint 2003: 154).

Anti-United Nations sentiment helps explain why the IPCC gains no respect from people holding this world view. With regard to the Kyoto Protocol, Flint writes 'The Kyoto Protocol is obviously another elite passion' (2003: 175) and proceeds to quote the Lavoisier Group, and prominent sceptics Fred Singer, Ian Castles and others on climate change—as well Brian Fisher of the Australian Bureau of Agricultural and Resource Economics (ABARE) at the time, whose economic modelling underpinned much of the argument about the potential severe damage to the Australian economy if Australia signed any significant Kyoto targets.

Propaganda techniques have been successfully adapted to much of what we understand today as public relations and marketing techniques. Common propaganda techniques that can be recognised in the metaphors and language applied to 'sceptical' climate change discussions include: the use of fear, name calling, glittering generalities, euphemisms (that appear as metaphorical language or misleading labels) and appeals to what the Americans call 'plain folks' and Australians would call 'the battlers' (Delwiche 1995). An extension of these techniques can be seen in the public relations advice on how to frame climate change science to stress uncertainty offered by Frank Luntz (2003). In this framing, specialist climate scientists can be regarded as just another 'academic elite', out to keep their jobs and their perks.

That public relations advice has had an increasingly strong hold on what passes as news can be shown from media research. Modern politicians and business leaders all have public relations advisers. By 2001 researchers were asking 'how it is that practitioners of public relations have managed to usurp authorship of the news?' (Ward 2001: 178).

Talkback rules

While public relations techniques help mould what audience perceive to be reality, in Australia considerable influence in the public war of ideas is waged by talkback radio. Radio professional John Faine argues that a major strategic advantage for the Howard federal Coalition view of the world from the mid-1990s was the understanding of the importance of talkback radio. 'Talk radio has overtaken all of the forms of media—electronic or print—as a political medium in Australia. It has become the daily agenda setter and the preferred organ for national and state leaders to sell policies and ideas' (Faine 2005: 167).

Consequently, the commercial radio talkback hosts, along with the politicians they interview, wield immense power on the daily issue agenda for public discussion. Action on climate change has not been a favoured item. The most popular talk show hosts in major Australian cities make their bread and butter by taking extreme positions, saying outrageous and abusive things about politicians or climate scientists, among others, and diving right into the economy versus the environment divide of the culture wars.

One-time Labor political adviser and self-styled 'left' media commentator Dennis Glover, described the role of the media, whether radio, press or television, in 1990s agenda-setting:

> (Prime Minister) John Howard had the powerful levers of government at his disposal to influence public opinion, but he had something more—a strong forward pack of media supporters willing to pick up a policy or a message, and smash through the opposing teams defences ... the screaming front page 'exclusives', rabid opinion columns, unbalanced editorials, soft radio interviews and opponents made timid by their own ethical codes, must be close to what the Italian political theorist Antonio Gramsci had in mind when he coined the term 'hegemony'. (Glover 2005: 213)

Dominating the channels of information in the way I have discussed in this chapter helped install an ideological hegemony that obscured earlier knowledge of what James Baker, former head of the US National Oceanic and Atmospheric Administration (NOAA), said of climate change science in 1997: 'There's no better scientific consensus on any other issue I know—except perhaps Newton's second law of dynamics' (Gelbspan 2005: 73).

A majority of media coverage during the second half of the 1990s and in subsequent years ignored both this message and acres of previous newsprint that took human agency as a given, to report instead that human involvement in climate change was contested and prompt action was not going to happen because it would cost the economy and cost every family.

The fascinating issue that arises is the nature of perceived 'reality' to which the public reacts on a daily basis. In this story it has changed from one decade to the next.

Academic research, particularly from science history, neuroscience and psychology shows, along with the documentary evidence trail, that what we know as reality can be, and is, manipulated by elite agenda setters within societies. The media and politicians, reflecting ideas, values and influential backers, set and reset the agenda. Their beliefs and policies dictate what the public hears, aided by the heavily researched tools of public relations, the basing of news on what politicians say, and media practices that favour drama and controversy.

But what about scientists themselves, what role did they play in the revised story of 'can't do' on climate change action that gripped Australia and other countries in the 1990s?

8. Influences on a changed story and the new normal: scientists' beliefs and public scepticism

Sceptics—look at their track record; for an important group of sceptics their primary qualification is geology; in this argument, that offers red herrings and doesn't help the policy process.

Geoff Love, Director, (Australian) Bureau of Meteorology, speaking at the 5th world conference of science journalists, Melbourne, 18 April 2007

Climate scientists are a very small cabal that actually don't study climate change, they study weather change … but the expert group of scientists on climate change … is the people you've just referred to, geologists.

Bob Carter, marine geologist, speaking on *Mornings with Paul Murray*, 6PR, Perth, 11 March 2011; quoted on ABC *Media watch*, 21 March 2011

Scientists too have values and beliefs. The world views of different scientific disciplines can significantly influence science and society discussions like climate change.

Different disciplinary groups act like academic tribes, with their own set of intellectual values and their own patch of cognitive territory (Becher 1994). Armed with this understanding, it's easier to grasp the challenges faced by the sprawling, multi-disciplinary task of unravelling climate change and also understand where some of the staunchest sceptics have come from. Climate science has required that scientists from a wide range of earth and environmental sciences learn to cooperate, and to accept each others' data, often for the first time, to affect the progress that has been made.

Policy gridlock is not unusual for controversial environmental science research. A recent report looked at the interaction of disciplinary differences with social values and 'normative lenses' (i.e., what a discipline considers *should* be the case) and found:

> In areas as diverse as climate change, nuclear waste disposal, endangered species and biodiversity … and agricultural biotechnology, the growth of considerable bodies of scientific knowledge, created especially to resolve political dispute and for effective decision-making, has often been accompanied instead by growing political controversy and gridlock. Science typically lies at the centre of the debate, where those who

advocate some line of action are likely to claim a scientific justification for their position, while those opposing the action will either invoke scientific uncertainty or competing scientific results to support their opposition. (Sarewitz 2004: 386)

So it's worth taking a closer look at the beliefs of several disciplines that have been at the forefront of sceptical debate about anthropogenic climate change.

Whether or not one calls economics a science, its normative (i.e., what *should be*) assumptions and theories have exerted a significant influence on the public discussion and on political attitudes towards climate science in recent decades. Several other disciplines feature prominently as well. Many of Australia's most oft-quoted climate change sceptics, with seemingly relevant scientific credentials, are either geologists—for example Bob Carter (James Cook University), Ian Plimer (University of Adelaide and mining company director)—or they are meteorologists or climatologists, particularly William Kininmonth (former administrator of the Bureau of Meteorology's (BOM) National Climate Centre) and, for a while, Brian Tucker (after leaving the CSIRO Division of Atmospheric Research, which he headed at the time).[1]

The term 'sceptic' (or 'denier' or 'contrarian') is a common label for those who reject the Intergovernmental Panel on Climate Change (IPCC) assessments on anthropogenic climate change or deny human agency in the phenomenon. While I use the term sceptic, as defined above, I concede that this use is problematical for scientists, who would typically characterise themselves as sceptical by training and inclination.

Disciplinary differences help explain the tenacity of some sceptics who are not necessarily linked to corporate special interests and whose continuing public debate in the face of overwhelming evidence may appear puzzling. While the general public and many journalists may think that anyone called a climatologist or a meteorologist or geologist must be an expert on climate change, and some may be, the disciplinary assumptions of these professions, particularly from training dating back 30–50 years or more, are different from that of today's specialised atmospheric and climate scientists.

Eminent biologist Peter Doherty notes that the complexity of modern interdisciplinary science, which collates a huge amount of interrelated data from

---

1 Two of the leading 1990s US sceptics invited to Australia, Patrick Michaels and Robert Balling, were climatologists by training or employment (*Sourcewatch*). Displaying their own disciplinary perspectives, statisticians like Ian Castles and Bjorn Lomborg have entered the debate with a sceptical point of view. Clive Hamilton in his book *Scorcher* (2007) discusses the Australian sceptics, as does the website *Sourcewatch* which provides backgrounds on prominent sceptics.

various disciplines, has left 'a few old geology and meteorology practitioners, in particular, very uncomfortable with this process and [they] over-state the case that their "historical knowledge" is being ignored' (Doherty 2009: 9).

Geologists, climatologists and meteorologists have been taught that past or present conditions are the only valid predictors of weather, climate or future planetary situations. In this view, modelling data of future events can always be attacked as weak and unsubstantiated. These disciplinary backgrounds would incline the practitioners to promote a natural variation explanation and reject human activities as causing climate change phenomena.

'Balance of nature' global change and geology

A detailed history of the discovery of global warming/climate change by US physicist and science historian Spencer Weart shows that, until recently, those studying earth processes held an implicit belief that there is a 'balance of nature' that would correct any disturbances created by human activity. Indeed, a central idea was that human activity is insignificant against the great planetary forces that shape and reshape our world.

Calculations made since the late 1800s about the heat-holding significance of a rise in CO_2 levels in the atmosphere drew arguments that there exist compensating or balancing mechanisms, such as increased cloud formation. Weart commented:

> These objections conformed to a view of the natural world that was so widespread that most people thought of it as plain common sense. In this view the way cloudiness rose or fell to stabilize temperature, or the way oceans maintained a fixed level of gases in the atmosphere were examples of a universal principle: the Balance of Nature. Hardly anyone imagined that human actions, so puny among the vast natural powers, could upset the balance that governed the planet as a whole …. This view of nature—suprahuman, benevolent and inherently stable— lay deep in most human cultures. It was traditionally tied up with a religious faith in the God-given order of the universe. (Weart 2004: 8–9)[2]

While by the mid 20th century everyone also knew that there could be pivotal global changes such as ice ages—in fact the exploration of ice ages started

---

2   Lynn White Jr's seminal 1960s study *The historical roots of our ecologic crisis* set the stage for environmental history studies that acknowledged the ingrained Christian beliefs in Western culture about the roles of God, humans and nature.

climate change studies—the assumption was that this only happened on vast timescales, not on human time scales. So it was believed there was no immediate worry about any potential climatic changes.

Geologists were at the forefront of mapping out the ice ages, which brought them into climate studies, and their basic disciplinary assumptions conformed to the so-called 'uniformitarian principle'—that the present is always representative of the past. 'The principle was cherished by geologists as the very foundation of their science, for how could you study anything scientifically unless the rules stayed the same?' (Weart 2004: 9–10)

It is not surprising, then, to learn that Ian Plimer appeared on the fundamentalist free market Institute of Public Affairs (IPA) website in 2007 with a review article entitled *The past is the key to the present: greenhouse and icehouse over time* (Plimer 2007).

The firm belief that answers can only be derived from on-ground review of past earth history may have stemmed from a painful dispute among geologists that disengaged them from the 'catastrophist' legends of global change preserved in religious traditions, such as Noah's flood. Given this background, geologists were not about to entertain new theories of rapid, catastrophic change without a battle.

Science historian Naomi Oreskes provides a fascinating history of geological disputes over continental drift. Uniformitarianism was the geologists' answer to dealing with physical evidence balanced against the 'eighteenth century association, particularly in England, of geology with theology in general and with scriptural exegesis in particular'. Given this background, sudden or dramatic change, 'unaccounted for by the normal processes of daily geological life were all too close to miracles for most geologists' comfort' (Oreskes 1999: 204). At the same time, many scientists, like other members of society, privately retained a view that the world was governed by a 'normal' God-given order.

Normality and consistency: the bedrock of weather reporting

Beliefs in normality and consistency also pervaded the fields of climatology and meteorology. If one thinks about how the weather is still reported, even in a country as manifestly variable as Australia, it is in deviations from some hypothetical or statistical norm.

The science of climatology has traditionally been based on averaging seasonal temperatures and rainfall in the belief that statistics of the past 100 years, since records began, could reliably predict future decades. In this view, 'climate' equals a set of weather data averaged over the ups and downs. Principal clients have been farmers and engineers, who needed statistics to decide on crop plantings and to prepare for 100-year floods. While climatologists predicted seasons, meteorologists were using similar means to look at the next day's weather by looking at the recent past.

All three of these disciplines had developed a culture of relatively narrow, on-ground measuring and comparison that viewed modelling and theorising outside the box as problematic territory (Weart 2004). Brian Tucker, on retirement from the CSIRO, provided sceptical analyses for the IPA as a senior research fellow emphasising uncertainty and caution. In a letter following an interview he wrote: 'although perceptions of possible climate change depend almost entirely on numerical climate modelling, model results are generally accepted uncritically, with little cognizance given to the weaknesses inherent in model specifications, the mathematics used and the poor precision of model results.'

Tucker and other scientists sceptical of the science and policy debate over climate change have not appeared as ready to apply similar criticisms to economic modelling and its assumptions. Thus, in a critical piece written for the IPA, Tucker quoted at length from an economic analysis produced for the Electricity Supply Association of Australia in August 1994. This analysis predicted 50 and 60 per cent increases in energy prices and the elimination of the aluminium industry if the 1990 interim national emission reduction target of more than 20 per cent by the year 2000 went ahead. This economic modelling was accepted at face value, while climate science based on modelling was uncertain. In any case, it was Tucker's view that any impacts would occur slowly over centuries and that the policy response, therefore, verged on unnecessary panic that would just hurt the economy (Tucker 1994).

In Australia, the influence of these disciplinary positions can be seen also in the relative absence of the Bureau of Meteorology (BOM) from the evidence on the public record during the 1990s and the eventual emergence of Kininmonth as a prominent sceptic following his retirement. John Zillman, director of BOM from 1978 to 2003 was engaged with climate science policy advice to government, but the record indicates that this was mainly confined to acting through the processes of the World Meteorological Organisation and the IPCC. He was described as 'quite conservative' about climate science in a detailed 2004 article on the Australian sceptics and the Lavoisier Group (Fyfe 2004). The article quotes him as saying he is now convinced of the mainstream science of climate change and human agency, although he would not have been '10 years ago'— that is, in the mid-1990s.

Zillman's former colleague Kininmonth became active after his retirement in 1998 and promoted the view that climate change is a purely natural variation that takes place over long time spans and that human impact is minimal. This is consistent with the belief that past cycles always inform the present. Kininmonth told Australian Broadcasting Corporation (ABC) science program *Catalyst* in 2005 that:

> the science underpinning the greenhouse scenario is flawed. The computer models are at a rudimentary state of development. The actual science of climate would suggest that we are near the peak of global warming and that the prospect is in fact, in the longer term we're talking now thousands, to tens of thousand of years, is a gradual cooling. (Horstman 2005)

CSIRO atmospheric scientist Graeme Pearman, who since the 1980s had communicated the risks of loading the dice for climate change, retorted on the same program: 'I think it's rubbish. I think he's not an expert, he hasn't tested his ideas in the open literature, that's what scientists have to do.'

Like other sceptics in retirement from active science, Kininmonth has 'found fame in the twilight of his career,' noted Melbourne *Age* journalist Melissa Fyfe, in her article on the Lavoisier Group which promoted Kininmonth's thinking. He was also named as a science adviser, along with Bob Carter, at the US Science and Public Policy Institute. British professional sceptic Christopher Monckton has been the chief policy adviser for this sceptics organisation dedicated to 'sound science including climate scepticism' (http://scienceandpublicpolicy.org).[3]

Targeted attacks on environmental science, political interference

Disciplinary differences may have predisposed some scientists to a sceptical and combative stance in the public discussion on climate change. But the evidence shows there has also been a targeted attack on climate science, and environmental science generally, coming from some political and corporate interests, which have drawn on the same sceptic names. In the United States, in particular, there is evidence of political interference and political attacks on climate scientists and their data in the early 2000s under the neo-conservative government of George W. Bush.

---

3 When Kininmonth's book denying anthropogenic climate change was launched by the Lavoisier Group in 2004, Zillman agreed to launch it and then gave a remarkable speech supporting freedom of debate but criticising Kininmonth's non-peer reviewed analysis that assisted those who denied the anthropogenic influence (along with natural variation) when the preponderance of the evidence now pointed to it (Zillman 2004).

In March 2004 the US Union of Concerned Scientists (UCS) published an open letter called *Scientific integrity in policymaking* signed by 62 prominent scientists, including Nobel laureates, and heads of federal agencies and universities (Union of Concerned Scientists 2004). According to the UCS website, 12,000 scientists are said to have signed this document by 2010. The letter said that the Bush administration in the United States (2001–2009) encouraged systematic interference and misrepresentation of findings, including those on climate change, and that this compromised the integrity of science communication.

The letter spoke of 'consistent misrepresentation of the findings from the National Academy of Sciences, government agencies and the expert community at large'. The UCS also asserted that this misrepresentation was accompanied by 'disreputable and fringe science reports and [by] preventing informed discussion on the issue' (Union of Concerned Scientists 2004).

A 2007 survey of working government scientists in the United States supported these findings with personal testimony.[4] There are documented complaints about government reports being shelved, conclusions being altered or deleted, political operatives second-guessing scientists and cases of scientists being harassed by Congressional committees. The survey found that more than 40 per cent of respondents reported pressure to eliminate words like climate change and edit reports to change their meaning. Other practices that were reported included: not issuing press releases, changing press releases by injecting uncertainty or making communication so bland or technical that nobody would give it a second glance (*Atmosphere of pressure* 2007).

There is not a similar body of evidence of this level of interference in Australia. But a dampening effect on communication to the public can be assumed given charges of government scientists being 'muzzled' from 1996, when the conservative parties came to power federally (Cohen 2006; Pockley 2007; Hamilton & Maddison 2007). With the closely allied political and economic views of Australia and the United States through the latter part of the 1990s and early 2000s, a similar dampening approach to the science was on the cards.

Thus veteran science writer Peter Pockley described in an interview what the Coalition government under John Howard signalled to public science agencies: 'Scientists were told you don't say anything that might embarrass the government or the minister.' Control was also exercised through an increasing emphasis on commercialisation within the CSIRO and a de-emphasis through budget cuts on public interest science agencies, such as Atmospheric Research and Wildlife

---

4   The report *Atmosphere of pressure: political interference in federal climate science* was published by two non-government agencies: the Government Accountability Project (GAP) and Union of Concerned Scientists (USC). The UCS and GAP surveyed almost 300 scientists, carried out 40 interviews and searched thousands of agency documents (*Atmosphere of pressure* 2007).

and Ecology. 'These divisions were the target for political pressure during the 10 years of the Coalition … with the extraordinary notion that scientists have nothing to do with policy in these areas of climate and natural resources.'

The bigger picture shows a considerable body of documentation, largely outside the academic journals, about what some call a 'war' on environmental science that started in the United States during the 1970s and, coinciding with the 30-year neo-classical economic experiment begun under President Ronald Reagan in the United States, that gained traction in Australia as I have shown (Mooney 2005). That world view pits public interest in the natural environment against economic interests and 'the market', unless it is a case of using the natural environment for gain.

Climate sceptic ties to neo-conservative think tanks

In 2008, a trio of US social and political scientists published the extent of the links between scientific climate change sceptics and free market, neo-conservative think tanks (called neo-liberal in Australia). They found that more than 92 per cent of sceptical books published in the United States were linked to conservative think tanks and that 90 per cent of conservative think tanks interested in environmental issues took a sceptical approach to the evidence for climate change.

They concluded that the framing by sceptics of themselves versus the science was often not neutral but was 'organised by core actors within the conservative movement.' Promoting scepticism is a key tactic of the anti-environmental counter movement coordinated by conservative think tanks designed specifically to undermine the environmental movement's efforts to legitimise its claims via science (Jacques, Dunlap & Freeman 2008).

Their studies supported the view that conservative think tanks are politically powerful and funded by wealthy foundations and corporations. This tactic has been wielded successfully since the public battles over tobacco smoking and the hole in the ozone layer (Beder 2000; Rampton & Stauber 2002; Mooney 2005, Oreskes & Conway 2012). It's easy to forget that before anthropogenic climate change became a debate, a similar battle raged for 10 years from the mid-1970s over accepting human responsibility for the hole in the ozone layer.

The evidence shows that there is a close ideological affinity between free market, conservative North American (including Canadian) think tanks and those in Australia, such as the IPA, which receives considerable funding from the resource sector and is closely allied with the conservative parties. The same holds for the Lavoisier Group—which was established specifically, under the lead of mining sector representatives, to counter climate change science. Their ideas are amplified by mass media, particularly the News Limited media.

Think tank, political and media ties were boldly on display in the lead-up to the 2013 federal election. On 4 April 2013 *The Australian*'s publisher and long-serving IPA director and News Corp boss Rupert Murdoch was the keynote speaker at the think tank's 70th anniversary dinner. This was also attended by federal Coalition leader Tony Abbott and the world's richest woman, mining magnate Gina Rinehart. It is not unrelated that the Murdoch press in Australia after 2010 started a no-holds-barred campaign to oust the incumbent Labor government and install the conservative Coalition.

A few days later, Rinehart reported Murdoch's address in glowing terms in the conservative magazine *Quadrant*. She wrote that Baroness (Margaret) Thatcher would have applauded his thoughts and reported that Murdoch described his father as one of the postwar founders of the IPA in a proud battle against socialism, which is still being waged today. Their efforts helped to 'open up Australia by deregulating, privatising, reducing tariffs and floating the dollar'. Rinehart described Murdoch's assertion that Australians must be brought to understand that markets are not only efficient, but fair and moral agents bringing freedom and prosperity. In sum, Australia should unleash more economic rationalist policies (Rinehart 2013).

As a prime ministerial contender, Tony Abbott vowed to do so on the same night singling out environmental protections and climate change responses and bureaucracies in particular to be abolished along with a federal mining tax.

On that night the IPA's corporate and political soulmates were bold and revealing of the vision and agenda of this war of ideas. In this realm of thinking, environmental protection and in particular climate change science, and policy response, are counter to the interests of market forces and business empire builders like Murdoch and Rinehart.

The following example from the *IPA Review* is typical of how these battles have been waged over the years as an attack on the scientists seen to be standing in the way of 'freedom and prosperity'. *IPA Review* editor Mike Nahan wrote in an article entitled 'The demise of science':

> Why have so many scientists succumbed to being myth-makers? One answer is money. Shock and horror not only sells newspapers and generates donations for NGOs, it also generates funding for research. And as Professor Bob Carter discusses in 'Science is Not Consensus'

(pages 11–13) changes to the funding of science in recent years have increased the incentive for scientists to join in the doom and gloom. (Nahan 2003)

With Abbott in the prime minister's chair in 2014, the success of this ideological attack is manifest. *The Canberra Times* reported in an editorial bemoaning the 'puerile' debate about climate change: 'The newly elected Abbott government won office on the back of opposing even a modest penalty or price on carbon pollution, a policy that once had bi-partisan support' [and] 'Mr Abbott's chief business adviser, Maurice Newman publicly and without embarrassment labels climate change a "scientific delusion" ... he even says the carbon tax, which has been in place for just $1\frac{1}{2}$ years helped destroy manufacturing in this country' ('Self-interest key to weather debate' 2014).

Look up Maurice Newman and you find a free marketeer who has wielded influence with both sides of federal politics in the last two decades. A former stockbroker and investment banker, Newman served as chairman of the Australian Stock Exchange and later was appointed under the Howard government first as a director and then chair of the public broadcasting ABC board for much of the 2000s. Newman was a co-founder of the conservative think tank the Centre for Independent Studies (CIS), which promotes the late economist Milton Friedman's free enterprise vision. Newman has not been convinced by climate change science or the need for renewable energy. Under his watch at the national broadcaster, Murdoch delivered the 2008 Boyer Lecture on the theme 'A golden age of freedom'. Also in 2008 the ABC broadcast *The Great Global Warming Swindle*, a British program criticised for its inaccuracies by the British broadcasting regulator.

Global aids to uncertainty and inaction

Since the early 1990s, the fossil fuel lobby has mounted an extremely effective campaign of deception and disinformation designed to persuade policymakers, the press, and the public that the issue of climate change is stuck in scientific uncertainty. (Gelbspan 2004: 40)

The same distortions of the public discussion on climate change have been in effect in other Western English-speaking countries and perhaps not coincidentally where News Corp operates. But News Corp has also had a lot of like-minded help to influence the public, coming from the Global Climate Coalition and its constituent corporations. *New Scientist* reported that framing techniques, used in Australia and the United States during the 1990s and since, were also operational in Britain in the 2000s, featuring familiar global mentors.

In an editorial 'Still in a mess over climate change' (2006), *New Scientist* echoed what some environmental groups and investigators have reported since the 1990s about the oil company Exxon Mobil's long-standing and extensive funding of lobby groups, think tanks and individuals that, the science magazine said, misinform the public on climate change (Examples of Exxon Mobil's influence have been documented by Greenpeace's Exxonsecrets at www.greenpeace.org; Beder 2000; Gelbspan 2004; Mooney 2005).

New Scientist reported charges against Exxon Mobil that stemmed from no less than the Royal Society in London which sent 'a measured complaint' to the oil company about these practices, only to be ignored. *New Scientist* fumed that such arrogance towards one of the world's oldest scientific institutions 'seems to rival their contempt for good science itself' ('Still in a mess over climate change' 2006: 5). The editorial described public discussion in Britain in the mid-2000s as beset by familiar public relations and propaganda tactics of sowing confusion and name-calling, (e.g.: theories of climate change being described as the 'big lie').

Further evidence for the corporate strategy of using sceptical scientists to sow uncertainty into the public discourse surfaced in the Australian media in 2007 and again involved Exxon Mobil. Following the release of the fourth IPCC assessment, *The Sydney Morning Herald* revealed in a page one report that Exxon Mobil was offering $10,000 to scientists to dispute the IPCC findings ('Bribes for experts to dispute UN study' 2007).

Recently, evidence has emerged from New Zealand of an organised attack, with international links, orchestrated to sow doubt on scientific weather readings. The New Zealand Climate Science Coalition is allied with the International Climate Science Coalition (whose science adviser is none other than Australian geologist Bob Carter, according to its website). It has waged a three-year court battle to discredit temperature data gathered by the National Institute of Water and Atmospheric Research (NIWA). According to newspaper reports, taxpayers have been saddled with the court costs as the Coalition went into liquidation rather than pay costs.

The data showed that at seven stations from Auckland to Dunedin, between 1990 and 2008, there was a warming trend of 0.91 °C, according to NIWA scientist Jim Salinger, who reported the saga in January 2014 (Salinger 2014).

The language of negative framing

Calling the global scientific consensus a 'debate' has been one of the most familiar framing devices of the past 20 years and reflects a recognisable propaganda technique: to encourage emotional reactions, in this case of uncertainty, in an audience. Decrying those who want to 'shut down debate' or 'politicise the science', or are said to merely seek self-advancement/research funding and are therefore indulging in public 'scaremongering', have been frequent examples of stirring the pot of uncertainty.

Of course, even a cursory look at mass media outlets, the talkback radio sphere or the blogosphere shows that negative labelling techniques are not confined to anti-climate science and free market intellectual warriors linked to right-wing think tanks. Public discussion has been polluted on a wide front with language that encourages negative emotions and distrust of scientific expertise and political leadership, arguably leading to more public disengagement and confusion.

The combative propaganda approach serves a political economy focused on free enterprise rights and a re-emergent vision (as spelled out by Murdoch in his IPA speech) that society need be no more than a collection of self-interested individuals engaged to markets. Taken together, these forces are a recipe for more 'whistling in the dark' as land, sea and atmospheric systems are damaged, linked to a growing catalogue of climate-driven natural disasters.

Ingredients for cooking up confusion and distrust of science

In the mid-1990s conservative members of the US Congress charged that environmental science was biased and a congressional committee investigated—putting on the record a catalogue of techniques still used today by those who oppose public interest science. The congressional report—*Environmental science under siege: fringe science and the 104th congress*—documented attacks against climate scientists and others working with environmental and public health regulations (Brown 1996).

The congressmen who made the allegations and the sceptic scientists they called to testify alleged that environmental scientists couldn't be trusted. The investigating committee found there was no basis to this and made a useful summary of how public communication has been confused as a result of attempts to discredit the scientists.

The recipe for confusion used by sceptical individuals and organisations included: abusing the goodwill of democratic and scientific practice by diverting time and attention to the views of a few dissenting scientists; launching sceptical attacks that consistently mix scientific data, opinion and policy advice while mainstream scientists do not have this luxury; dismissing or misusing scientific conventions, including peer review, consensus and uncertainty. In the 15 years since, it is possible that not just climate scientists, but aspects of science itself, have come under siege as a narrative of mistrust is elaborated.

A year after the Brown report, a concerted attack on environmentalists and climate change science appeared on mainstream British TV station Channel 4 and was repeated by the ABC in Australia in 1998. The 1997 two-hour *Against Nature* documentary directed by Martin Durkin made good use of a number of familiar media techniques. Durkin included opinions from alleged expert scientists without providing context or biographical information to prove their claims to be climate change experts. He also failed to mention the larger context or balance of evidence—the IPCC international scientific reviews that have sounded the alarm since 1990.

The presenters interviewed included S. Fred Singer, a retired US space physicist and science administrator, who was active in the battle to deny industrial responsibility for ozone depletion before moving on to anthropogenic climate change. He is a good example of a professional contrarian and was amongst the sceptics invited to present their views on climate change in Australia during the mid- to late 1990s, sponsored by conservative, free market think tanks but also by the CSIRO (Beder 2000). As is often the case with retired sceptics, Singer did not publish his critiques in peer-reviewed science publications. His activities through various campaigns against public interest science are instructive.

In the 1990s, Exxon Mobil supported Singer's policy research group and he first earned his sceptic tag by appearing as an expert for the tobacco industry. The tobacco campaign, as many researchers have pointed out, honed the public relations and communication strategies that are still used by corporations today. That campaign popularised the phrase 'junk science', which subsequently has been used by both sides to discredit opposing views. It also popularised the use of scientific experts rather than corporate spokespeople to make the industry's case (Beder 2000; Rampton & Stauber 2002).[5]

The UK Independent Television Commission subsequently found that *Against Nature* was misleading and distorted interview information. Investigations found that, while the script echoed extreme conservative arguments against environmentalists, the program's director described himself as a Marxist. He was linked with a small far-left group, the Revolutionary Communist Party, which also had links to several of the featured experts. This group believes that sustainability or environmental concerns are conspiracies against progress for Third World people (Monbiot 1997). So it seems attacks on environmental science come from multiple political standpoints within Western society.

---

5 Singer was also a lead author of the highly criticised but effective *Leipzig Declaration on Global Climate Change* that showcased dissenting opinions (Beder 1997: 238).

A follow-up program, *The Great Global Warming Swindle*, was made by the same director and UK Channel 4 and aired in Australia in 2008. It yielded similar complaints about distortions and inaccuracies of the science, which were found to be valid by the British broadcasting regulator (Cubby 2008). When shown on ABC television in Australia it received a large audience. The slick packaging, persuasive images and experts with science labels, led to anecdotal responses that this program successfully created public confusion or uncertainty.

British journalist George Monbiot was asked in 1997, after the first *Against Nature* program, which he investigated, how ideas like the ones showcased in that program could receive a two-hour, prime-time television slot. He said he had found that:

> Many television executives hate environmentalism. They see it as a grim *memento mori* at the bottom of the pictures, spoiling the good news about cars, clothes and consumerism. So when the film-makers suggested an all-out assault on environmentalists, their proposal fell on fertile ground. (Monbiot 1997: 1)

Fostering a climate of uncertainty since the mid-1990s has been an effective strategy for creating and maintaining public confusion and official inaction on climate change. It's worth taking a closer look at why we expect certainty and how applying uncertainty to environmental research findings has been a favoured tool for contesting research findings about future consequences of human activities. There are lessons for controversial environmental science and public policy throughout.

9. In search of certainty and applying uncertainty

I am not a scientist but it's always seemed to me that one of the strongest arguments about climate change is that ... if I were to say to you that there was a 60 per cent chance your house might burn down, you would take some insurance.

British Prime Minister David Cameron, speaking in the wake of typhoon and tidal surge Hainan that killed thousands in the Philippines and flooded Vietnam ('British PM's warning on climate', *Canberra Times*, 17 November 2013)

When advising politicians, the scientific community has devoted considerable attention to uncertainties, and has sought to adopt a position of 'objective neutrality' that has allowed advocacy groups with vested interests to dominate the advice on options for policy response.

H. Hengeveld, '1994–95 in review: an assessment of new developments relevant to the science of climate change', *Climate Change Newsletter* (DPIE and BRS) 1997

There is ample evidence that scientists, journalists and policymakers expressed 'certainty' in the early 1990s about how humans are warming the planet by producing excess greenhouse gases with industrial and consumer activities. This was a key driver of the political and public understanding exhibited between 1987 and 1992.

The evidence then shows that the language of scientific certainty not only changed, but that uncertainty was also deliberately constructed to throw doubt on the scientific conclusions. In the 1990s this occurred in all the Western, English-speaking democracies and globally to the extent that organisations like the corporate Global Climate Coalition and sympathetic news media had influence.

In the United States, where climate change discussions closely resembled those in Australia, atmospheric scientist James Hansen (who has spoken out about the risks of anthropogenic climate change over the course of two decades) addressed the US Congress in 1988 about the certainties. He said he was 99 per cent certain that global warming had begun, based on the series of warm years in the 1980s.

As reported by Robert M. White, then president of the US National Academy of Engineering, 'The public took notice. His opinion prompted Congress to consider whether the prudent course was to move rapidly to *legislation* aimed at preserving the habitability of the planet from catastrophic consequences' (White 1990: 22). I emphasise legislation in this quote because it is indicative of the times, when regulation to lower public risk was not considered politically out of bounds.

In his comprehensive review of US climate change public communication up to mid-1990, White recounts that internationally, political leaders took action between 1988 and 1990 based on the certainty of scientists' language and also thanks to the widespread drought at the time, which greatly focused attention on the issue. White's historical account of the issues and influences in the early 1990s is more significant because his own reaction was caution about accepting the science, therefore he cannot be accused of being an activist for mitigating emissions. He writes that after the politicians got on board with the climate science, the counter-framing started and sceptics in the scientific community joined open debate in the pages of *Science* on the certainty and validity of climate science.

Discussion in the public arena at that time expressed no doubt that the energy economy of current civilisation was the issue, along with land and water use and the effects of population increase. While sympathetic to the sceptics, White had no quarrel with the concept of human agency, saying that this understanding has deep historical roots within science. He also describes intervention in the policy process by advocates of uncertainty in the form of sceptics Richard Lindzen and Frederick Seitz who, together with a long-range weather forecasting expert and several other scientists, wrote to President George H.W. Bush questioning the science and urging research rather than action: 'Thus the great climate debate had been joined' (White 1990: 22).

A 2004 study by US science historian Naomi Oreskes tested the argument that behind the conventional scientific language, couched in degrees of certainty or uncertainty, there is no published consensus on the occurrence of anthropogenic climate change. Oreskes showed this is not so.[1] She analysed 928 abstracts published in the refereed literature between 1993 and 2003, and listed in the ISI database with the keywords 'global climate change'. She found no disagreement and wrote:

> Scientists publishing in the peer-reviewed literature agree with the IPCC, the National Academy of Sciences, and the public statements of their professional societies. … Politicians, economists, journalists and others may have the impression of confusion, disagreement, or discord among climate scientists, but that impression is incorrect. (Oreskes 2004a: 1)

1   Oreskes told a reporter she decided to test the consensus after her hairdresser said *she* wasn't worried about global warming because scientists 'don't know what's going on'.

Five years later, a survey published by Peter Doran and Maggie Zimmerman of Earth and Environmental Sciences at the University of Illinois came to a similar conclusion. They canvassed more than 10,000 earth scientists and received responses from 3,146. Of these, more than 97 per cent of the specialists on the subject (i.e., 'respondents who listed climate science as their area of expertise and who also have published more than 50% of their recent peer-reviewed papers on the subject of climate change') agreed that human activity is 'a significant contributing factor in changing mean global temperatures' (Doran & Zimmerman 2009: 23).

The researchers commented: 'It seems that the debate on the authenticity of global warming and the role played by human activity is largely nonexistent among those who understand the nuances and scientific basis of long-term climate processes'. Relevant to the disciplinary beliefs of some geologists and meteorologists (discussed in a previous chapter), they found that these two fields had the lowest 'yes' response to the question about whether human impact is changing mean global temperature (47 per cent and 64 per cent respectively, compared with 82 per cent for the group as a whole.)

The demand for unarguable 'proof' of climate change arises both from a misunderstanding of scientific conventions and the deliberate deployment of uncertainty by critics. Since environmental impacts often raise costs for some sectors that can influence the policy process, demanding proof of impacts has long been a convenient stalling mechanism.

Certainty and the Rachel Carson case

The historical record shows that no major scientific shifts in understanding on how the planet works (e.g., the once radical and contested evidence for continental drift or ice ages) have waited for absolute proof. More recent history shows the certainty debate about climate change bears a startling similarity in creating controversy to marine biologist Rachel Carson's seminal 1962 popular science book *Silent Spring*. But the political response at the time to Carson's work was also significantly different to the response I have outlined for climate change, and is worth noting as another way to look at the uncertainties of predicting environmental impacts.

Carson's work was about the effect of pesticides on biological systems, including humans. The book was the first to give the public insight into the potentially disastrous effects when synthetic chemicals enter the environment and about environmental pollution generally. At the time Carson was fiercely attacked (an onslaught that continues to this day on contrarian websites). Among other things she was charged with engendering fear (therefore charged as being emotional and unscientific) about the consequences if humans continued down a reckless path *vis a vis* the natural environment (Oreskes 2004b).

For the scientists who attacked Carson and her work in the 1960s, doing good for humanity was a major theme. Carson's opposition came principally from industries that made or relied on environmental chemicals, including the food industry and their related scientists. Some had ties to the pesticide industry, but others had beliefs and values dedicated to a large and inexpensive food supply, which was challenged by Carson's work and ecological concerns.

Similarly, some of the more aggressive but persuasive arguments for demanding proof and certainty from climate change science have come from people committed to equity and 'progress' for developing societies through fossil fuel energy technologies. In Australia, the mental acrobatics between the desire to lower greenhouse gases while being the world's largest coal exporter is often justified by talk about helping China and other developing countries attain Western living standards—a value that often accompanies other, deeply held ideals like 'freedom' and democracy.

Carson's classic case highlights the difficulties that continue to face many environmental scientists dealing with complex and evolving systems in a scientific culture that is used to counting and measuring. Similarly to climate change research, and before that the decade of industry resistance to evidence about the hole in the ozone layer, Carson's work had measurable evidence lagging behind modelling and proxy data, leaving the door open to arguments about certainty and proof.

But on the political front and with regard to respecting risk, the Carson story is notably different from the contemporary response to climate change. Carson was supported by a 1962 President's Science Advisory Committee (PSAC) review (under President John F. Kennedy). The review was brief and easy to read and acknowledged the trade-offs of all human activities, but concluded on balance that more harm than good might be the long-term outcome of pesticides for plants and animals including humans.

> PSAC never claimed that the hazards of persistent pesticides were 'proven', 'demonstrated', 'certain', or even well understood; they simply concluded that the available data were adequate to show that harms were occurring, warranting changes in the pattern of pesticide use. (Oreskes 2004b: 374)

The panel took seriously suggestions for alternative technologies, did not accuse Carson of hidden agendas, and did not use uncertainty as a justification for inaction. Perhaps most significantly, PSAC placed the burden of proof and certainty not on the public interest scientists but on the emitters or polluters, in this case, those who argued that persistent pesticides were safe. The climate change analogy is that those who argue that humans are *not* having an impact would face the burden of proof.

How science-speak contributed to uncertainty and delay

During the 1990s and since, most specialist climate change scientists returned to the scientific conventions of cautious communication, often stressing uncertainties, after a brief period characterised by plain English communication with the public. Journalists were not the only ones to comment on this. It was noted as early as 1989 that there was a disjunct between scientific and public understanding of percentage uncertainties. Climatologist Ann Henderson-Sellers told a reporter:

> There was a big furore in the USA during the past year when a scientist told a Senate enquiry, he was 99 per cent sure that the Greenhouse Effect was with us now. Unfortunately, a number of my colleagues disagreed with him because they're only 95 per cent sure, and the media had fun with that. Yet when I surveyed a number of people about what level of confidence they wanted from the scientific community—before they'd start planning for the future—the answer was 50 per cent. (McKenzie 1989: 34)

In 1997, Canadian researcher Henry Hengeveld examined the contribution of scientists' own style of communication in promoting the confused public discussion that had taken hold by the mid-1990s and has continued since. He reviewed 885 papers published on climate change in 1994–1995 and noted the effects on policy. A report on his work appeared in the federal government's *Climate Change Newsletter*. He wrote:

> Although misinformation spread by self-interest groups is a factor, the scientific community has been ineffective in communicating its information and concerns in a manner useful and comprehensible to lay audiences. Furthermore, when advising politicians, the scientific community has devoted considerable attention to uncertainties, and has sought to adopt a position of 'objective neutrality' that has allowed advocacy groups with vested interests to dominate the advice on options for policy response. Some authors have suggested that scientists should take a more proactive role as policy advisors, while in Australia pretty well the opposite happened over the 1990s. (Hengeveld 1997: 21)

Melbourne *Age* journalist Geoff Strong has pondered why it has taken so long for the scientific messages about climate change to push through to real action. He decided that a problem is the scientific definitions of uncertainty/certainty, which sound like hedging to the general public and to their elected and often scientifically ignorant representatives. Strong has noted that he was reporting on the greenhouse phenomenon 20 years ago and that some scientists emphasised the uncertainties even then, and continued to do so during subsequent years. He says they wrote in terms like:

> Well we are not 100 per cent certain but ... in science-speak, that means they could have been 95 to 99 per cent certain but were leaving the 1 per cent margin for error in case somebody ripped them apart in a scientific paper ... The world's greatest gamblers, the insurance industry, didn't need that level of certainty. It had been banking on scenarios being right since at least 1995. (Strong 2005: 1)[2]

CSIRO Division of Atmospheric Science former administrator and communicator Willem Bouma told Strong that, in hindsight, perhaps scientists should have worded their predictions differently and conveyed more certainty because two decades have since been lost. Strong commented: 'By appearing uncertain, they might have protected their backsides, but allowed a whole army of vested interest groups such as the fossil-fuel lobby and right-wing think tanks to attempt to lever apart the argument and create 20 years of delay' (Strong 2005: 2).

The evidence from government, business and other public documents, and several hundred popular news articles (see chapter 4) clearly shows that in contrast to the later science-speak, a certainty of language framed the discussion about climate change up till 1992. Further evidence can be drawn from the following examples from the language characterising the 1990 Intergovernmental Panel on Climate Change (IPCC) report, in comparison to later reports in the 1990s and since.

The 1990 IPCC report

The first, 1990, IPCC report has been all but forgotten in contemporary discussion of the IPCC assessments that are delivered to national governments every five to six years. Two things stand out from the 1990 report in comparison to the 1995 and 2001 reports, and compared with the knowledge of the 2007 reports. Firstly, the 1990 report confirms that the basic findings of science and impacts hardly

---

2   The quoted article by Strong found its way, via the internet, to a climate change blog—desmogblog.com ('we're here to clear the PR pollution that clouds the science on climate change'). The writer, Jim Hoggan, contrasts Strong's article to the US *Cape Cod Times* for 30 October 2005, which, he says, provides a perfect example of why climate change deniers are still in there with a fighting chance. The *Times* lauds a climatologist for perfect integrity 'the absolute insistence on total scientific certainty' (desmogblog.com 2005).

changed during the following decades. Secondly, the communications style and language were clear and definite—in marked contrast to later reports where the language was marked by uncertainty and, worse, reverted to disciplinary jargon and technical detail.

The 1990 report is characterised by a notable level of plain English. The scientific assessment report executive summary starts: [my emphasis in italics]

We are *certain* of the following:

There is a natural greenhouse effect which already keeps the Earth warmer than it would otherwise be.

Emissions *resulting from human activity* are substantially increasing the atmospheric concentrations of the greenhouse gases: carbon dioxide, methane, chlorofluorocarbons (CFCs) and nitrous oxide. These increases *will enhance the greenhouse effect*, resulting on average in an additional warming of the Earth's surface. The main greenhouse gas, water vapour, will increase in response to global warming *and further enhance it*.

We calculate *with confidence* that:

inter alia

[Impacts] Under a business as usual scenario a global mean temperature increase *of about 0.3 degrees C per decade* with an uncertainty range of 0.2–0.5 degrees C—this is greater than that seen over the past 10,000 years. With controls under different scenarios, the rates of increase could drop by 1/2 or 2/3.

The authors admit many *uncertainties in predictions* of timing, magnitude and regional patterns due to incomplete scientific factors such as sources and sinks, clouds, oceans polar ice sheets. (Houghton, Jenkins & Ephraums 1990: xi)

The introduction that follows this executive summary again speaks plainly and with confidence, which makes the document accessible to a politician, journalist or other lay reader. This first IPCC scientific report was described by its chairman, John Houghton, as the work of 'most of the active scientists working in the field. Some 170 scientists from 25 countries have contributed either through international workshops or written contributions'. A further 200 scientists were involved in the peer review of the draft report (Houghton, Jenkins & Ephraums 1990: v). It therefore summarised the known body of research at the time and felt able to report with certainty.

Houghton acknowledges minority opinions, but says the peer review of the draft report helped to ensure a high degree of consensus amongst authors and reviewers of the information presented (and presumably of the language used to communicate). 'Thus the assessment is an authoritative statement of the views of the international scientific community at this time.' This foreword, written in July 1990, concludes on a hopeful note, lauding: 'a significant step forward in meeting what is potentially the greatest global environmental challenge facing mankind' (1990: v, vi).

1989–1992: government reports showed little doubt

Even without the IPCC report, expert advice had convinced Australian Government deliberations by the late 1980s. In December 1989, an inquiry by the Senate Standing Committee on Industry, Science and Technology showed its certainty and looked at ways to reduce the impact of the greenhouse effect with these words:

> The experts with whom the Committee met confirmed that there is irrefutable scientific evidence that the composition of the atmosphere has been, and continues to be, altered significantly by human activity.
>
> There is the risk that if the response to this problem is delayed until the evidence of significant climatic change is irrefutable, it may be too late to avoid some of the more extreme changes that could occur … slowing and reversing the changes in the atmosphere will be slow and difficult. Consequently, it is essential that an early start be made in implementing changes. (Senate Standing Committee on Industry, Science and Technology 1989: 1)

In its own words, this Senate committee accepted the scientific evidence of atmospheric change, and that it was induced by humans. It did not require 100 per cent measured certainty of climate change in order to take action, which was understood to become more costly with delay. The committee communicated all this in certain language.

In 1989 the Australia and New Zealand Environment Council (ANZEC), in an agenda item on the draft National Greenhouse Strategy, urged all state governments 'as a matter of priority to pursue all available measures to reduce greenhouse gas emissions' (ANZEC 1989). As we have seen, by 1990 many states had developed response plans.

Also in 1989, the Labor government under Bob Hawke released a state of the environment report wherein a response to anthropogenic climate change featured prominently. 'Significant climate change ... would have major ramifications for human survival' (Hawke 1989: 28). This document (which also ushered in the era of Landcare and tree planting) agreed that waiting for 'conclusive scientific evidence' was not necessary, but an early start on action was. A year later, in October 1990, the federal government drafted its interim planning target to reduce greenhouse gas emissions by 20 per cent (from 1988 levels) by the year 2005 (Commonwealth 1990).

As late as 1992, as the Rio Earth Summit introduced the UN Framework Convention on Climate Change to drive the international agenda on this topic, one continues to find certainty of language in Australian federal government documents. For example, a 1992 federal government *Climate Change Newsletter* confirms that the discussion had gone well beyond debating whether the greenhouse phenomenon exists or not and was dealing with the emission reduction targets. Significantly, this newsletter acknowledges that energy demand management (i.e., efficiency) could make a major contribution to achieving the government's reduction targets (Department of Primary Industries and Energy 1992).

At the time, the newsletter was edited by the Department of Primary Industries Climate Change Group, where it remained until late in the 1990s, produced by the Bureau of Rural Resources. The change in its language in the 1990s is telling. In the early years the language is certain and direct and frames do-able responses that were later reframed as being unacceptable or undo-able.

The February 1992 newsletter's lead article, for example, outlines the government's ESD (ecologically sustainable development) Greenhouse Working Group report to the Department of Arts, Sports, Environment and Territories with the good news that energy services could be upheld while emissions were reduced with innovative management of both supply and demand.

The sticking point was likely the next sentence 'but that achievement would require high levels of government intervention' (Kretschmer 1992: 1). The high level of government 'intervention' (the market term for regulation or incentives)—principally through managing demand by mandating efficiency, fuel substitution and urban planning—would become unacceptable, ideologically, during the remainder of the 1990s. But while it was a real option, it shows how certainty appeared in tandem with strong policy response.

Uncertainty marks the reframed story after 1992

To assume that the economy and the environment were basically opposed, an old tradition in Australia, offered a valuable tool for those who framed Australia's 'national interests' as synonymous with the existing energy and export system. It helped to characterise the environmental science and talk of risks as 'uncertain', thereby neutralising a challenge to the economic policy agenda. Marginalising environmental scientists and advocates helped as well.

Energy consultant Alan Pears advised the Victorian Government on changing its energy policies before the politics in that state shifted to economic rationalism by 1992. He saw how discrediting the science proved a powerful weapon to keep energy supply unchanged in the 1990s. Economic modelling on costs and manufactured uncertainty about the science reframed the public story by the mid-1990s to a focus on jobs and costs. 'By 1994 ABARE had convinced the Department of Energy with its [economic] modelling,' he said.

Pears also saw 1992 as a tipping point—away from policy progress on climate change action backed by public knowledge, definite communication and positive leadership. This switch correlated with a change of government in Victoria and also with the ascent of Paul Keating to prime minister. Public messages became framed as 'any action is going to hurt', while competition policy and deregulation guided energy sector 'reform'.

Lobbyists win with job talk while 'degrees of uncertainty' lose

At the federal level by the mid-1990s, former policy adviser Sue Salmon saw firsthand the weak position of the Department of Environment, which was the conduit for the science. She told me, 'there was a whole lot of that "bring in a sceptic" strategy and it was understood that public confusion made it easier to continue with business as usual.'

She also recalls the strong presence of lobbyists from the coal and paper industries. 'Their message was effective and powerful. It was about income and jobs while we were talking about degrees of uncertainty.' These trends only intensified with the marginalisation of the environment movement under Prime Minister John Howard (1996 on) and the lack of interest in what the science, including the IPCC, said.

The media then amplified for the public the climate of uncertainty coming from politicians and this was topped by the more deliberate campaign of scepticism encouraged by News Limited and the right wing think tanks. Claire Miller, a journalist who was then working at *The Age*, a Fairfax publication, remembered the uncertainty that crept into Australia's politician-driven news system:

> The question was, is it real or scaremongering? Legitimacy comes when the government is taking it seriously. Under Hawke it was big profile. Under Keating it went back to a junior ministry; the media follows what politicians are talking about so then politicians stopped talking about it and the media stopped too; meanwhile the community thinks it is being 'fixed'.

Retreat from definite language becomes the norm

By 1995, the IPCC too had started to mute its language with jargon, highlighted uncertainty and retreated to technical discussions in the summary for policymakers. The late US atmospheric scientist and IPCC member Stephen Schneider told me that commentators from the (anti-greenhouse science) Global Climate Coalition were pressuring IPCC members after the first, 1990 IPCC report. At the same time, politicians got more involved in the reporting process. It was a far cry from the 1988 Toronto conference of scientists and government officials that led to the establishment of national emission reduction targets in the first place.

At that conference, Schneider convinced delegates that plain and forceful communication was the way to go. Then *Sydney Morning Herald* journalist Leigh Dayton, backgrounding the 1995 IPCC report, painted a vivid picture about the urgency to communicate that emerged at the Toronto conference:

> The anxious experts feared that if human beings continue to load the atmosphere with heat-trapping greenhouse gases like carbon dioxide— produced largely by burning coal, oil and wood—the world would be doomed to an 'impending crisis' of unbridled climate change: global warming, increased storms and droughts, sea-level rises and other extreme and hard-to-predict weather events, not to mention the human chaos and suffering that would ensue.

> But what could a group of scientists, administrators and environmental hangers-on do? 'Give the public and politicians firm answers, not

statements of scientific uncertainty,' vehemently argued one young turk, Dr Stephen Schneider, now a leading climate modeller at Stanford University in California.

And so they did. To this day the final statement from that extraordinary meeting remains one of the most unnerving scientific pronouncements ever made: 'Humanity is conducting an enormous, unintended, globally pervasive experiment whose ultimate consequences could be second only to a global nuclear war'; it is 'imperative to act now'. (Dayton 1995a: 29)[3]

A few years later Schneider had changed his mind. Looking back in 2007, he cited the pressure from corporations in oil, coal and gas on the international IPCC process following the plain English 1990 report. In response he said he was the driving force behind correlating the terms 'likely' and 'very likely' to percentages of certainty in subsequent IPCC reports in an effort to standardise the language. 'Did it work?', I asked him. He answered:

Well it worked for scientists. Not sure what the public got out of it. But I believed that the public would settle for lower percentages if framed by credible scientists. Credibility of the scientists is key. (However) there is the related problem of scientists not drawing conclusions under the framework of not overstepping the policy line because of politicians' censure.

IPCC reports from the mid-1990s, cautious and technical

So we see that from the 1995 report on (compared to 1990), language in the IPCC science summary for policymakers (which may be the only document most politicians and journalists read or are briefed on) became more diffuse and technical and open to interpretation. In 1995, while saying the 1990 predictions and scenarios had held, the science summary is considerably less to-the-point than in 1990 and also focused on measurement or quantifying impacts. Here's a taste. The reader must get to page four before learning:

---

3   Leigh Dayton and Gavin Gilchrist, her colleague at *The Sydney Morning Herald*, wrote a number of detailed and unequivocal articles in the months following the 1995 release of the 2nd IPCC assessment report—outlining the extreme weather and other risks posed by ongoing global warming and climate change. These articles, like others from the period from the *Herald* and the *Age*, provide an excellent historical record showing that many environmental or science reporters remained certain of the problem's existence and that most of what is currently understood about climate change was understood then as well—despite the possible communication barriers of a less than user-friendly IPCC report.

The balance of evidence suggests a discernible human influence on global climate; any human-induced effect on climate will be superimposed on the background 'noise' of natural climate variability ... our ability to *quantify* the human influence on global climate is currently limited because the expected signal is still emerging from the noise of natural variability, and because there are uncertainties in key factors. These include the magnitude and patterns of long term natural variability and the time-evolving pattern of forcing by, and response to, changes in the concentrations of greenhouse gases and aerosols, and land surface changes. (IPCC assessments 1995, Working Group I, pp. 4–5)

The communication effort is not helped by the 1995 summary for policymakers on potential social and economic responses, which reads like an academic economics treatise, perhaps reflecting its authors' disciplines (Lee & Haites 1996). It sends no urgent signals and may well have remained unread by policymakers because of its style. Retreating into the difficult, technical and inaccessible was another safe strategy to avert criticism, along with stressing uncertainty.

Such a deliberate strategy to sow disinterest or uncertainty, was identified by two US science policy pressure groups: the Government Accountability Project (GAP) and the Union of Concerned Scientists (USC). The report, *Atmosphere of pressure: political interference in federal climate science* (2007), looked at tactics during the George W. Bush administration in US federally-funded departments and agencies dealing with resource and environmental matters.

The 2001 IPCC science report continues with a technical style of language and delivery. The summary for policymakers announces within the second paragraph that it describes the current state of understanding of the climate system and 'its projected future evolution and their uncertainties' (IPCC assessments, 2001, Working Group I, p. 2). It lays out its 'judgmental estimates of confidence' along the likely, very likely continuum suggested by Schneider. The reader is told that it is 'very likely' (which the scientists equated to a high 90–95 per cent chance) that the 1990s was the warmest decade and 1998 the warmest year in the instrumental record since 1861. Proxy record data going back thousands of years is 'likely' to be certain, which translated still allows a range up to 90 per cent certainty.

It is not hard to see the uncertainty and desire for another opinion that this language is liable to cause in a lay audience, let alone an unfriendly policy audience. In 2001 the anthropogenic or human influence is described as: 'The influence of external factors on climate can be broadly compared using the concept of "radiative forcing"', footnoted with a technical explanation.

What the 1990 report called 'emissions resulting from human activities that are increasing concentrations of greenhouse gases' is called 'increased concentrations of atmospheric constituents' in 2001. It is not until the second last page of the executive summary that the reader learns human activities have continued to increase greenhouses gases and 'their radiative forcing' since the 1995 report and that this is due to fossil fuel burning and 'land-use changes'—a benign-sounding jargon term that mostly refers to deforestation.

By this time the federal government's *Climate Change Newsletter* had also largely retreated into technical reports compared with its earlier direct and accessible news reports. The overall picture is that a focus on measurement and technical, quantified reporting became the yardstick of credibility and also a way to justify 'go slow' as the 1990s turned into the 2000s.

How that scientific uncertainty frame was interpreted can be seen from an exchange between scientist/science communicator Tim Flannery and journalist Tony Jones on the Australian Broadcasting Corporation's (ABC) *Lateline* in May 2007. Flannery was asked to comment on a sceptical documentary. He said the documentary did not reflect the consensus of scientists globally, namely: that it was '90 per cent certain' that human activity produced the increased greenhouse gases leading to global warming that caused climate change. Phrasing it this way, caused Jones to respond 'yes but, that means there is 10 per cent uncertainty, which surely leaves an opening for this sceptic debate?'.

Risk assessments conservative

While at CSIRO, earth systems scientist Michael Raupach, (now heading the ANU Climate Change Institute), commented in an interview in 2008 on the language of uncertainty and its effects at the IPCC at that time and on domestic response:

> The sceptics [internationally] have been very active in throwing 'sand in the gears' causing the IPCC to use very carefully calibrated language. Lots of people including me think that this has led IPCC statements to fall on the conservative side. Likewise, sceptics have been very influential on our government so scientists have had to moderate their language to reach people in policy.

> [In so doing] CSIRO has been pulling its punches on climate change. Especially in the mitigation area we have failed at plucking the real 'low-hanging fruit' such as energy efficiency. We are not doing anywhere near what is needed. [At the same time] sceptics like the Lavoisier Group

have wedged open and magnified uncertainties, but only on one side; they stress the possibility that climate science may overstate the threat, but ignore the equal possibility that climate science is understating it.

Raupach also offered a scientist's unvarnished view of how governments had failed to grasp the necessity for steady emission cuts and said: 'all this talk about cap and trade [emissions trading], nuclear and clean coal, is just "greenwash" to avoid confronting the need for real strategies for rapid, sustained reductions in fossil fuels.' In 2014, Raupach said:

> Much has changed since 2008. In Australia, climate science is under renewed and coordinated attack from three directions simultaneously: the fossil fuel industry, much of government, and sections of the media. This means that there is now an even stronger need than in the past to defend evidence-based science against anti-science attacks. The recent (2013–2014) IPCC Fifth Assessment provides a detailed, authoritative and fundamentally challenging account of what the realities are: the emission reduction rates that are needed to stay below a warming target of 2 degrees above preindustrial temperatures are very, very steep, and their direct policy implications are enough to draw the triple fire attack.

Greatly assisting the climate of uncertainty was the media trend to frame climate change as debate, opinion and in need of balance, as I have shown. There was also a trend to treat each IPCC report as if discovering for the very first time that humans were causing global warming. For example a 2001 *Sydney Morning Herald* story is typical of the coverage. In 'Six degrees hotter: global climate alarm bells ring louder', we learn that 'World temperatures may increase by as much as six degrees Celsius over the next century, leading climate change scientists say in an alarming report that adds new urgency to the warnings on global warming (Schauble 2001). … *And for the first time scientists agreed that the warming is mostly due to human activity*' (my italics).

This 'just discovered' human agency hook can be identified in media reports on successive IPCC assessments. It shows how the story is reframed in the media and it also shows how reporters dispense with context and background.

Atmosphere of pressure on media

As this uncertain framework unfolded, journalists who remembered the history, understood the science and reported it in plain English, found themselves under pressure. Geoff Strong remembered: 'I was taken to the Press Council in 1999 by a reader for writing about global warming a decade on. My alleged crime was I hadn't given oxygen to those who didn't believe.' Former ABC environmental journalist Alan Tate, who says the ABC was 'completely supportive' during most of the 1990s of reporting on climate change, recalled that his bosses there were inundated with emails from sections of corporate Australia decrying that coverage and calling for his sacking.

Tate said the strategy he saw coming from industry complainants was to 'sow doubt about the climate science'. This was done through seminars, forums, and climate sceptics. 'The coal industry and Rio Tinto had the ear of the Prime Minister and the Canberra press gallery and [most of] corporate Australia was disengaged until after 2000 … while the green movement was still heavily focused on forests and also disengaged.' He believed that this lead to 'a completely confused public discussion'. By the time he left the ABC in 1998 the 'deep uncertainties idea' had settled with the editors at the national broadcaster.

Murray Hogarth, another environmental journalist active in the 1990s, said it was easy to find an opposing point of view and that often there was a problem with finding Australian scientists willing to be quoted at all, or quoted in a simple and understandable way. In this way, public information was nudged towards more manufactured balance that became a normal part of reporting on climate change, and balance was hedged by uncertainty.

Institutional change silenced scientists

Part of the problem of finding scientists willing to communicate as the 1990s unfolded was the chilling effect of institutional changes on scientists' ability to publicly communicate the consequences of climate change on society. This came to be seen as commenting on government policy and it was forbidden. As early as 1987, change was affecting the major scientific body involved in Australian atmospheric research. An October 1987 government internal memo on climate change work at CSIRO sounded the alarm saying restructuring might require finding a 'major funding sponsor'; that is, government financial support would wane (Department of Arts, Sports, the Environment, Tourism and Territories 1987).

In the view of former federal politician Bob Chynoweth, who was on the advisory board of the Division of Atmospheric Research at the time, the public interest science in the CSIRO was gradually 'squeezed down' as the organisation was reorganised. So too was government-funded scientists' ability to communicate freely, particularly from 1996 on with the Howard Coalition government.

A scientist who never stopped reporting on climate change, Griffith University professor Ian Lowe, wrote what others had been saying privately: that the CSIRO under former Chief Executive Geoff Garrett during the Howard years developed 'a culture of managerialism so wary of offending government, that scientists have been instructed not to comment on issues that have policy implications. Even within universities ... there is now increasing pressure to conform' in the face of a disapproving government that controls the purse strings (Lowe 2007: 60–61).

As the organisation was restructured to serve the needs of industry, CSIRO climate change and other environmental researchers came to face a double barrier: a government employer who discouraged scientists from talking about impacts of climate change, along with energy and resource industry 'partners', many of whom came to sit on the CSIRO Board and on cooperative research centre (CRC) and flagship boards (Pearse 2007).

Graeme Pearman told me that the defined role of the CSIRO changed and became a directive 'to build wealth' at the expense of sharing with society the outcomes of public-good research. He was in a position to observe this as chief of the Division of Atmospheric Research for 10 years. Veteran science journalist Peter Pockley agreed. 'A policy line is set, often on the basis of ideology or whim, and science is effectively urged to get on board the policy bandwagon ... it has taken four reports and 15 years to say what people like Graeme Pearman were saying in 1990' (Pockley 2007: 31).

John Williams, former chief of CSIRO Land and Water, says in the same 2007 article: 'we must get around the view that there is a clear definition between science and policy. It's nonsense to say that presentation of scientific information is a form of advocacy which must be avoided.'

The long-term effect was that those who agreed with the government policy position felt free to speak out while those who did not were intimidated into silence (Lowe 2007: 61). Pearman told Lowe in an account of the organisational trouble he encountered in 2004, while still a prominent member of the CSIRO Division of Atmospheric Research:

> As a climate scientist, I might inform [media] that the lifetime of carbon dioxide in the atmosphere means that the only way of stabilising global climate is by reducing emissions by 50 per cent by 2050 and by 80 per cent by 2100. In the current environment, that is seen as commenting on government policy of not setting reduction targets. (Lowe 2007: 63)

Reaping the longer term impact of these stifling trends that started in the later 1990s, Pearman found himself in trouble in 2004, when a report was released by The Climate Group, a business scientific alliance, convened by the insurance company IAG and the World Wide Fund for Nature. It involved Pearman for scientific advice and included some of Australia's major corporations outside the mining and resource sector. The report brought together evidence that climate change was starting to affect Australia.

In 2006, the ABC *Four Corners* television program, 'The greenhouse mafia', reported that Pearman came under CSIRO administrative pressure as a result of his work with The Climate Group. In the television program, reporter Janine Cohen asked, 'Talking about the need for a reduction in emissions and how much would be a safe level, is that really government policy? Isn't it about good science?' Pearman said, 'Well, I believe it is ... for 30 years all I've tried to do is convey to the community and to sectors of the community what good science suggests is the way forward' (Cohen 2006: 7).

Pearman says he was subsequently made redundant by the CSIRO, in the division that he had led as chief from 1992–2002. His communication work, together with that of scientific colleagues with the groundbreaking greenhouse conferences in 1987 and 1988 and definite public communication thereafter, made a significant contribution to the early good public understanding of the risks inherent in climate change. Pearman, his colleague Barrie Pittock and also Lowe are among the small group of Australian scientists and technologists who have withstood the pressure from the mid-1990s into the 2000s and continued to speak out clearly and publicly about the risks of climate change. More have joined their ranks in recent years, indicating a hopeful change for environmental scientists' ability to publicly communicate.

The muzzling of research findings was not confined to climate change. Another public sector scientist has told of his experience dealing with the federal bureaucracy during the 1990s. He said analyses he was contracted to produce for the federal government on the likely environmental impacts of population growth were never published because, he believes, they did not give the desired answers in line with population growth policies.[4]

Public interest research goes quiet too

Communication restrictions accompanied a more fundamental redrawing during the late 1990s under the Howard government of what constitutes 'the public interest'. Research into renewable energy, integrated pest management, tropical rainforests and the Great Barrier Reef was defunded (Lowe 2007) along

---

4 A detailed account of this 1990s collision between science and official immigration policy can be found in Lowe (2007: 65–70).

with the wind-down of the former CSIRO Division of Wildlife and Ecology and eventually Land and Water. The federal government redirected funding into commercial pursuits, including research for the coal industry.

The government was steering a course that equated the good of the private sector with the good of the community, implying that there was no separate public sector interest. This is consistent with the ideology of economic rationalism and its influence on public policies and public narratives in Australia in the past 20 years.

The muzzling of scientific conclusions and the downgrading of public interest environmental research helped cement the dominant narrative of uncertainty and 'scientists can't agree' enveloping Australia after 1996. Uncertainty was also helped by feuding bureaucracies representing environment and industry/ trade, and by previously neutral media executives backing down in the face of persistent corporate complainants whenever reports linked climate change to on-ground weather impacts.

Combined, these influences resulted in a level of public confusion that paralysed further calls for action from the grassroots up while the path was cleared for business as usual: status quo polluting energy providers, inefficient industries and consumer products, government policies that favoured cars and roads over public transport, in sum a vision of growth that was guaranteed to increase emissions.

10. Dicing with the climate: how many more throws?

While the industries they represent don't always see eye to eye, there are three things they all agree on: greenhouse emission constraints must be stopped; if they can't be stopped they must be delayed for as long as possible; and if they can't be delayed they should be written so as to exempt us from paying.

Guy Pearse, writing about the industry lobbyists who succeeded in getting a changed climate change narrative from the mid-1990s on (*High and Dry*, 2007)

The Abbott government remains steadfast in its plans to remove the carbon tax—now at $24.15 a tonne—which has helped to make black coal-fired plants, in particular, more expensive.

Peter Hannam, 'Wind energy at record levels', *Canberra Times*, 6 May 2014

The Institute of Public Affairs is bringing together the biggest names in the climate change debate. Make a tax-deductible donation today to help the IPA publish a new book of research, Climate change: the facts 2014, and continue to influence the climate change debate in Australia.

thefacts2014.ipa.org.au/

The 'biggest names' on the free market Institute of Public Affairs (IPA) contributor list include a majority who have appeared in the pages of this book. The list is sceptic geologists Bob Carter and Ian Plimer, retired meteorologist William Kininmonth, US sceptic scientists Patrick Michaels and Richard Lindzen, UK sceptic and John Howard informant Nigel Lawson, Canadian-born libertarian writer Mark Steyn, former IPA environmental editor Jennifer Marohasy, News Limited columnist Andrew Bolt, and Liberal Party stalwart John Roskam.

Also contributing to the volume is IPA's deregulation expert and climate change sceptic Alan Moran, who has revolved through the Industry Commission, provided intellectual ammunition to deregulate and privatise Victoria's public assets in the early 1990s and then worked as that state's energy chief, helping to undo efficiency programs. This is a good indication that the same voices and influence from the 1990s are continuing the same battle against climate change public knowledge and effective response, now with a sympathetic federal government again on side, more radical than before.

Before the 2013 federal election, the IPA published its wishlist for an incoming government under Tony Abbott that would make 'Australia richer and more free'. Amongst the top 12 wishes were 'repeal the carbon tax and don't replace it; abolish the Department of Climate Change; abolish the Clean Energy Fund' (Roskam, Berg & Paterson). Amongst the first actions of the Abbott government were to move to do just that.

Meanwhile a climate report in early 2014 that hardly raised a blip in the national conversation was the news that the West Antarctic ice sheet had started to melt without possibility of reversal, promising to raise sea levels within a century by 1.2 metres (possibly, as the *New York Times* reported, '10 feet or more' in coming centuries) (Rignot et al. 2014). At that time a separate team from the University of Washington led by Ian Joughin published corroborating evidence in *Science*.

Putting it in the historical perspective of early scientific warnings of risk, the *New York Times* reported:

> The new finding appears to be the fulfillment of a prediction made in 1978 by an eminent glaciologist, John H. Mercer of the Ohio State University. He outlined the vulnerable nature of the West Antarctic ice sheet and warned that the rapid human-driven release of greenhouse gases posed 'a threat of disaster'. He was assailed at the time, but in recent years, scientists have been watching with growing concern as events have unfolded in much the way Dr. Mercer predicted. (He died in 1987.) (Gillis & Chang 2014)

Meanwhile, weather events in early 2014 saw catastrophic flooding in the Balkans, off the scale of previous recorded flooding, which led to the displacement of people on a scale previously seen only in the recent Balkan civil war, and, again, reports of terrifying bushfires in the western United States following a long drought.

In Australia, however, this and other recent extreme weather events have not deterred the new national government, elected in September 2013, from attempting to undo the modest record of the previous government to lower greenhouse gas emissions and thereby do something as a nation to lower the risk of even more extreme and unpredictable weather.

It is said that, in war, the worst acts often occur at the end, when the tide is about to turn and conflict ceases. Is this possibly an analogy for the present Australian governments' (including some state governments) cascade of destructive, backward steps on climate change response?

While a manufactured budget 'crisis' has held the country in thrall, the real looming crisis of bad weather and economic costs related to climate change is scarcely on the radar. The government narrative (echoed by the same media voices particularly in the Murdoch outlets that supported this version of reality in the last two decades) signals 'no worries'—extreme weather events are a mere inconvenience that always happens in Australia. Longer term or overseas weather disasters are not part of the government narrative.

The backroom influence is the same as it was from the later 1990s to 2007; that is, the Howard era, with government policy allied to the advice of the extractive industry-funded and sceptical IPA and other free market think tanks. The country is being steered backward to a policy baseline from the past decades with the same influences of economics, beliefs and values, many of the same players, and the same rhetorical strategies that reset the national narrative in the 1990s.

The same influences are being deployed at a time that public opinion, some media analysis and policy responses have slowly moved away from the hegemony of thinking established in the Howard years. A post-carbon energy alternative has grown in possibility and acceptance since 2007. While a detailed accounting of the political economy and communication history post 2001 is not within the scope of this book, it can be said that the drivers that established the 1990s national story on climate change and banished the earlier good public knowledge have remained through the course of the 2000s so far, and there is now an attempt to re-establish total dominance.

One observer and economist in a good position to know this, and who has spoken out publicly, is Bernie Fraser, governor of the Reserve Bank of Australia from 1989 to 1996. More recently he has served as the chair of the Climate Change Authority, which was slated for dismantling by the Abbott government. The authority, which includes business leaders, was tasked with providing independent advice on Australia's carbon price, emission reduction targets and other response initiatives.

Fraser told public audiences in 2014 that it is a 'safe bet' the Australian Government will not change its minimum emission reduction target from five to 15 per cent, as recommended by the authority and has characterised as a 'barren' toolkit the new policies to pay industries not to emit carbon pollution. Confirming the familiar territory underpinning these policies including attacks on the windfarm sector, he told a regional newspaper that:

> [The federal government's position] is being buttressed by their big business supporters—the Minerals Council, the conventional energy generators—who want to see things go on the way they are and take

advantage of all the coal that they have in the ground and make a profit on it ... . The mining companies generally are strong supporters of the government ideologically. They are on the same wavelength.

[Asked about the public mood that this strong pushback is encountering, Fraser said:]

We meet quite a lot of business people who are concerned about the implications of global warming. And there is quite a growing renewable industry, as in solar. There was in wind too but it has hit a brick wall for the time being [due to government attempts to roll back the renewable energy target for production]. There's continuing investment in energy efficiency improvements. There are actually substantial numbers of people out there who want to do things differently. (Goldie 2014)

It remains to be seen whether current political events constitute a 'last stand' of sorts for the broader fossil fuel economy as it dices with climate change.

Policy drivers remain the same

What is clear since the 1980s is that decisive leadership, positive or negative, including beliefs, style and institutional arrangements have largely dictated whether meaningful action regarding climate change is on the national radar for Western democracies dominated by the values and economics outlined in this book. The importance of leadership was evident again from the enthusiasm initially generated in 2007 by incoming Labor Prime Minister Kevin Rudd and his pledge to do something big about climate change response. He memorably called climate change 'the biggest moral challenge of our time', and the public grew hopeful.

When his signature response of an emissions trading scheme fell apart, it caused much public disillusionment. This failure, too, can be viewed in light of the influence of resource extraction and energy intensive industries that retained the upper hand for the status quo in the 2000s. The trading scheme was watered down to please industry to the point that it lost support from environmental-minded members of parliament, notably the Greens, and therefore lost the parliamentary vote. This was an important element leading to Rudd's famous overthrow as party leader in view of growing party fears of electoral defeat.

The minority Labor government under Julia Gillard that then took the reins went into coalition with the Greens and two independents to retain office, and together they crafted a suite of renewable energy response measures, offering seed finance and supporting innovation and research. A price (tax) on carbon seguing to an emissions trading scheme was the main market mechanism of this response approach. Among other initiatives, a groundbreaking federally funded 'solar cities' pilot program of public–private enterprise showed that major and rapid change was possible in urban environments.

Within four years these initiatives were starting to bear fruit, boosted also by the global market for renewable energy that was bringing down costs of solar and wind systems around the world. Australia by 2014 had one of the world's best uptakes of domestic solar panels. Since 2010, wind energy had also become a large component of bringing down demand for traditional coal-fired energy.

According to an April 2014 news report, wind energy's share of Australia's main electricity market jumped to record levels helping to curb demand for coal-fired electricity and thereby lowering emissions. Wind had increased market share to '4.6% [while] black coal fired plants ... continued to operate well short of capacity. Greenhouse gas emissions from the National Electricity Market for the month were 5.8 million tonnes lower than a year earlier, down 3.5 per cent.' (Hannam 2014).

Despite such tangible evidence of progress towards a post-carbon energy market, poor communication by Gillard's government almost never linked the renewable initiatives or the price on carbon pollution to the larger goal of combating the risk of climate change to everyone in society. Instead the public narrative stayed firmly with the hip-pocket 'cost' story that was so well-entrenched by this time.

This left Gillard and her government vulnerable to a singularly nasty in execution but successful fear campaign mounted by the opposition (now the government) around the theme of 'tax' and broken promises. The fear of a new tax was not diminished by the Gillard government's cash payments to citizens, which were supposed to offset the flow-through impacts of the 'pollution tax' on business. The defeat of that Labor–Green alliance in 2013 by the radical conservative Abbott government was at least partly due to the public's acceptance of the cost arguments of the anti-carbon tax campaign.

Lessons from the climate change culture wars

The history of the last 20 years of Australia's response to global warming and climate change is a tale of power, profit and eventual unwillingness to accept social and economic change, falling back on a suite of traditional beliefs and values signalled to the public through the strategic application of the English language.

With the loss of political leadership willing to respond rapidly and effectively to the inherent risks of climate change, as was present in the late 1980s and early 1990s, this history shows the triumph of those who want no change from 'business as usual' and who frame everything in terms of cost rather than risk management and ethical response.

It is also a lesson about the dominance of communication in setting a society's sense of what is 'real' and that this reality can be reframed within a few years by politicians and media working in tandem. In the course of this history, Australia also changed from having a social democratic political approach up to the early 1990s with an 'accord' between capital and labour and a more inclusive approach to decision-making.

In regard to climate change response, for a few years, this inclusive approach included scientists and environmental organisations as mainstream partners, and was implemented in a politically bipartisan manner for the public interest. The public was presented with a science story that would affect everyone without manipulated 'balance'. Regulation of markets for the public interest was still not a banned concept. Under those circumstances a social consensus on action was easier to reach.

A major change that affected both communication and the political landscape in the 1990s was the final cementing of free market 'economic rationalist' ideology as the way to look at the world, as the new reality. Citizens were now stand-alone consumers in a relationship with 'the market'.

In the absence of committed leadership and then with negative leadership, public interest communication about global warming and climate change now had to persuade each individual to look beyond a manufactured debate and a mounting campaign of uncertainty about the science and the need to lower emissions. The result which favoured the status quo energy economy was a fog of public confusion and disinterest.

Neo-classical economic ideology took over the long-standing growth and progress myths that inform industrial cultures. In Australia and elsewhere it also allied with other free enterprise values in post-colonial societies that placed the economy *versus* the environment, and assumed the need and right to exploit natural resources (like coal). It also tapped into other traditional beliefs about human exceptionalism from the natural environment as well as a modern belief in technology fixing everything eventually.

The triumph of belief over evidence in an increasingly fragmented information environment has been one significant outcome in Australia of these influences that were nailed down in the 1990s, along with the advent of new information avenues, particularly the internet.

The outcome has been that the possibility of citizen demand for real and effective response action was successfully neutralised for the past two decades. Similar leadership changes, economic theories, beliefs and values also persuaded other English-speaking democracies—United Kingdom, United States, Canada— to go slow on climate action to various extents. Responses from other nations have varied as international negotiations show, but it can safely be said that leadership, economic models and beliefs and values also dictate their responses. The net outcome domestically and internationally is that we have lost 25 years of potential response action.

Instead, emissions are still mushrooming globally and scientific warnings are increasing about escalating, even runaway, climate change impacts while the window of opportunity is rapidly closing to stabilise greenhouse gases at a level bearable by human civilisations.

What we heard before being persuaded not to

What can we take away from this story of a nation that buried its once good public understanding of global warming and related climate change response measures?

Firstly, it points to public communication and message-framing as key, because the scientific message about anthropogenic climate change, unembellished by political translation and reinterpretation, has remained consistent since the 1980s.

The message since the 1990 Intergovernmental Panel on Climate Change (IPCC) report told us that, from the time of the Industrial Revolution, the enhanced greenhouse effect, global warming and climate change have been building. This is thanks to human activity—primarily emissions from burning fossil fuels, but also emissions from agriculture and deforestation.

If not mitigated by drastic cuts in greenhouse gas emissions from industrial activities and by slowing the rate of vegetation clearance (a sink for greenhouse gases), more extreme weather patterns bringing drought, fire, flood, severe storms and sea-level rise will become catastrophic for human populations and other species. The analogy has repeatedly been drawn that humans are loading the dice for more extreme events.

The late 1980s and early 1990s was a period when environment and economics were reconciled, or at least a focus on 'ecologically sustainable development' attempted a holistic approach, to ultimately fail as economic interests were once again pitted against the environment and those that defended it.

Every climate change response we know of now was known 25–30 years ago, starting with energy efficiency, which if applied across both domestic and industrial sectors, at one time was considered adequate to cut emissions significantly. This was in accordance with the 1990 Australian greenhouse gas emission reduction target of stabilising emissions at 1988 levels by 2000 and reducing them by 20 per cent from that level by 2005. Every state had a plan. Renewable energy, transport, urban planning, retaining vegetation were all extensively suggested as options and related research was funded.

The story was framed as one of opportunity for new industry as well. Significantly, regulation of then state-owned energy markets to manage demand for efficient use was considered acceptable and was indeed underway—until it all went the other way with deregulation and competition policy.

Mass media reporting generally reflected the early good understanding and will to act, consistent with both the prominent role of scientists at the time and also the media formula of quoting what people in power say as the main news. Prime Minister Bob Hawke and his environment minister were on board, as were state governments.

As political leadership changed, however, and the political response to lowering emissions switched from 'can do' to 'can't do', the media swung with it, using the same narrative and language as the politicians. In Western democracies like Australia, politicians and media together have set the agenda, which has become particularly noticeable with a growing loss of media diversity.

Let's talk plainly about risk, and ethics

The public record, including extensive newspaper reports and government documents, shows that framing the public discussion of climate change around risk confronting all levels of society (whether the risk is economic or about health) was a key element of the early public knowledge and political will to act. A good comparison is the public's understanding of insurance risk.

Other elements were a commitment to global good citizenship and care for future generations that, for a while, enjoyed bipartisan political support (an impossible political goal nowadays?). In the early days, atmospheric research scientists led the discussion and were quoted in the media using direct and certain language, as did the first 1990 IPCC report. Scientists also worked effectively with policymakers and community groups at a time when they were not constrained from talking about the implications of their research—later characterised as talking about policy and not to be done. Scientific experts did not at first (as they did later) lead with uncertainties when talking to politicians or other lay audiences.

Cognitive linguistics explains that much of public communication is what people hear, not what is said to them, and that messages might have to be framed differently for different audiences. Thus scientific uncertainty translates to lay people as 'scientists don't agree', or 'don't know'. This field of knowledge also explains how people can be manipulated by language that appeals to their core values like family and nation, or not liking taxes, so that they ' hear' what politicians want them to hear.

When it all changed to a different reality

From the mid-1990s, the political narrative moved away from a story about risk and global good citizenship along with domestic opportunity for new industries and cost saving through efficiency. The story became a drama of national self-interest, said to be threatened by outside forces like the United Nations—explaining some of the disdain for the UN-sponsored IPCC reports. The national interest was framed as being synonymous with the interests of industries that extract mineral resources and fossil fuels, notably coal.

In tandem, the business press no longer freely criticised Australian industries' inefficiency as a major reason for Australia's emissions. The early reporting saw environment, technology and even political reporters adding the science in context. Such context went missing by 2001 as political and economic reporters confined themselves to what politicians were saying.

What they were saying by this time was that human agency was uncertain and, anyway, Australia was exceptional as the world's largest coal exporter, other mineral exporter and energy intensive hub for multinationals, such as the aluminium industry. The political frame had become that changing or regulating the energy economy to reduce our emissions was against the national interest and therefore against every family's interest.

In the mainstream press, balancing climate change reporting with sceptical voices became standard by 2001, and opinion pieces on the subject went up 10-fold from a decade earlier, adding to the impression that the subject was debatable and a matter of opinion and belief.

Economic modelling became the lens for policy after 1992. The cost to mainstream Australia of a threat to status quo energy industries—without addressing a balancing benefit to society of emission or pollution reduction, trumped the narrative of risk from green groups and scientists. These groups became framed as special interest and counter to the national interest. Any action had to be voluntary, cost-neutral and market-focused.

Australia was not alone as a Western democracy under the influence of neo-classical market economics experiencing a seismic shift away from an evidence-based, science-informed national stance on climate change that highlighted risks, responsibilities and new opportunities. The 1990s economic discourse was about how things *should be* in a globalised economy—soon repeated in media and policy documents as the way of the world.

Beliefs and values trump evidence

Unpacking what drove this shift in the 1990s we find politically powerful beliefs and values had reasserted themselves, along with lobbying from the resource, finance and agricultural sectors.

In Australia (as elsewhere and, it has been argued, particularly in 'pioneer' countries) there is a strong set of 'no limits' beliefs—including assumptions about progress, growth and the unquestioned benefit of developing natural resources (one belief is that they are limitless). These have become allied with beliefs in a saviour 'techno-fix' if there are environmental problems. Granted, science and technology have spurred these expectations since World War II in spectacular fashion.

People in Australia and other Western democracies are also subject to deeply held beliefs in human or Christian 'exceptionalism' from other species and the natural environment. That we are different and special appears self-evident to many people. Could we be immune to climate change? These values while always present, made a strong comeback in the 1990s in Australia under both major political parties, but particularly under the conservative parties after 1996.

Such underlying values were highly compatible with the neo-classical, economic rationalist ideology (Reaganomics or Thatcherism being other names for the same theory) that dominated Australian society by the mid-1990s. The mechanism for radical idea change was largely through communication in politics, media and institutions. Markets became accepted as entities that could rule society; individual enterprise was said to be superior to communal groupings or interests and the private sector as more effective than government in delivering services and deciding what is good for the nation.

With this world view, the public interest becomes indistinguishable from corporate interests (which were also granted individual rights) and citizens became consumers with individual rights, mainly to consume and vote in elections, but encouraged to lose a sense of communal interests and responsibility.

Public relations advice spelled out not only how to engage people's core values—about freedom, family, jobs, nation, growth, progress, but also how to mount a campaign focused on promoting uncertainty and the belief that 'scientists don't agree' to stop the public from demanding action on climate change—regardless of the consistent science message. And so it has happened.

Lobbyists from resource, energy-intensive and agricultural business sectors and market-oriented think tanks have had unparalleled access to government decision-makers in a revolving door of professional advisers. Great influence was also exerted by News Limited, with a virtual monopoly in Australian print media circulation. The Murdoch media shared the notion that accepting climate science is unwarranted and a threat to business and has spent the last 20 years conducting a 'culture war' on this issue. Through politics and media these reasserted beliefs and values had taken over the whole society by the early 2000s and have returned in force in 2014.

In the Fairfax press I looked at, while science and technology reporters continued to retell and update the original science story of risk, their voices were outnumbered by political/economic reporting and, to differing degrees, sceptical opinions that said the opposite. Columnists, think tank publications and talkback radio hosts lambasted climate scientists as being self-serving seekers of research grants.

The establishment of the internet from the early 1990s mainly had the effect, before the 2000s and the advent of social media, of enhancing the opinion versus fact juggling act. While it enhanced scientists' ability to correct misinformation and archive those facts for future reference, it also offered the same ability to sceptics who readily developed blogs with science names that allowed people to cement forever sceptical and outdated scientific arguments.

Audience fragmentation was one result that has made it harder to communicate science messages in an individualistic, market-driven culture. In terms of the analysis in this book, professional and conventional journalism practices have continued to hold sway over the dominant public narrative, regardless of the advent of digital media platforms. It remains to be seen to what extent digital and social media will boost and facilitate citizen demand for climate action.

The media's preference for adversarial drama (intensified by the entertainment needs of television and radio coverage) accepted the emerging public role of sceptic scientists, encouraged by conservative think tanks and corporations fearing loss.

Many public sceptics have been geologists, climatologists and meteorologists. Regardless of what the titles imply to the lay public, these disciplines are not synonymous with climate change specialist. Based on their training, however, many have been sincere in their doubt of model-driven climate science, stressing that past planetary experience is the key to future events and only on-ground measurement is relevant.

A US Congressional committee found that many public sceptics routinely played outside the system by not publishing for peer review, by abusing scientific conventions of courtesy and democracy and by mixing fact and opinion or policy recommendations in their statements. Media conventions of conflating all scientists to equal status of expertise ('scientists say') has aided the sceptics and further confused the discussion.

Muzzling government scientists from talking about consequences, under the banner of interfering in policy, has played a significant role since the later 1990s, along with the demise of much public interest research funding and indeed the notion of 'public interest' as a legitimate sector. The record reflects a research landscape dominated by successive corporate restructures at the CSIRO, with a mandate for industry-focused research and new internal guidelines discouraging scientists' from speaking publicly.

As the policy narrative has moved from a conviction that effective response was possible, and the science narrative has evolved from certain and direct language, a major US review of government climate science reports notes that use of the technical, obscure and difficult in scientific reports to the public and media has been a deliberate tactic by governments to generate delay, disinterest and inaction.

What Australia has experienced with the combined effect of these influences has been the successful manipulation of public reality to impose a climate of uncertainty about global warming and climate change risk. For the last two decades this has been matched by an absence of coordinated, effective policy response (with a similar dynamic apparent in other Western, English-speaking countries).

Those years are lost to effective action, while global greenhouse gas emissions continue to accumulate. Contemplating another 'hottest year on record', the science tells us that reversing (long term) the threat of more catastrophic weather outcomes becomes harder every year as a result of our recent history.

That same history, however, is also a roadmap of what people once heard and thought. Effective action on climate change will start when society decides that things can be handled differently, as they once were.

A chronology of some major climate science/policy milestones

1800–1910	Industrial Revolution; at the beginning of this period, level of CO_2* in the atmosphere is about 290 parts per million (ppm) according to the ice core record; technological advances include coal-fired energy with related emissions and means for expanding land clearing; sanitation and medical advances promote population growth. *This chronology refers to CO_2 (carbon dioxide) not CO_2-e (CO_2-equivalent, which includes other greenhouse gases), as the measure of greenhouse gas emissions.
1896	Following the work of John Tyndall in 1861 showing that diatomic molecules absorb infrared radiation, Swedish scientist Svante Arrhenius publishes first calculations that planetary temperatures depend on greenhouse gases, speculating that human activity burning fossil fuels creates 'extra' CO_2 that might make the earth's temperature rise significantly over time.
1939–1945	World War II; nations expand their mission begun in the 1920s to control and exploit world oil supplies, adding more emission sources. Following World War II and technological innovations, resource exploitation, forest clearing, and population expansion explode.
1930s	Scientists suggest anthropogenic global warming is underway driven by more CO_2 and other greenhouse gases in atmosphere due to human activities. This was known then, and until 1990s, as 'the greenhouse effect'.
1950s	With computer technology, scientific advances allow modelling of the atmosphere, and understanding of climate feedback that accelerates warming or cooling trends, plus the realisation that oceans would not be absorbing all the CO_2 produced by humans.
1960	Detection of annual rise of CO_2 in the atmosphere and measurement at 315 ppm.
1967–1968	Calculation that doubling CO_2 will raise temperatures by several degrees; understanding that polar ice sheets could collapse and elevate sea levels.

1970	First World Environment Day signals strong upsurge of environmental interest and understanding. In the United States the creation of the National Oceanic and Atmospheric Administration (NOAA) creates world's biggest funder of climate research. Scientists begin organising and disseminating risk messages about human impacts on climate.
1972	Further research of proxy records (ice cores mainly) confirm possibility of rapid climate change within a millennium (later brought down to decades).
1975	Discovery of damage to the ozone layer and the beginning of a 10-year battle for an international agreement to restrict human-induced causes is a precursor to global climate negotiations, with many of the same sceptics and societal challenges evident as would appear in responses to the theory of the greenhouse effect. That ozone-depleting chemicals and ozone itself can contribute to the greenhouse effect is shown in the next year.
1970s	Better understanding gained of other possible influences on climate, including sunspot and orbital cycles.
1979	Second oil 'energy crisis' results in an upsurge in renewable energy technology, efficiency measures, smaller cars, calls to lower consumption—showing the feasibility of these technologies and behavioural changes (this understanding and these technologies were still influential in the late 1980s). First report on the greenhouse effect by US National Academy of Sciences says it is 'highly credible' that doubling atmospheric CO_2 will raise average global temperatures by 1.5–4.5 °C; World Climate Research Program launched. Election of Ronald Reagan as US President (and Margaret Thatcher as UK Prime Minister) starts two decades of backlash against environmental understandings and activism. It has been noted that a related set of beliefs dominated Anglo/American countries—United Kingdom, United States, Canada, Australia: neo-liberal market ideologies underpinned by beliefs in limitless resources and a self-adjusting natural world.
1980	The Australian Academy of Science organises a conference to review the thinking of leading scientists on the greenhouse effect. Playboy magazine covers the threats posed by the greenhouse effect, extensively quoting Australian scientists.

1981	Scientific prediction is made that greenhouse warming 'signals' would emerge from background 'climate noise' by 2000 and be measurable; 1981 declared 'warmest year on record'.
1985	Villach, Austria: United Nations Environment Programme/ World Meteorological Organisation (UNEP/WMO) scientific conference yields major public pronouncement by scientists linking anthropogenic increases in greenhouse gases with global warming—showing consensus within climate science community and calling for international action to curb emissions; a 541-page report is produced in 1987. The conference statement acts as a catalyst for global action. It opens: 'As a result of the increasing concentrations of greenhouse gases, it is now believed that in the first half of the next century a rise of global mean temperature could occur which is greater than any in man's history.'
	Antarctic ice cores show that CO_2 and temperature went up and down together during the ice ages.
	Scientific calculation that disruption—with ice-melt fresh water—of the North Atlantic ocean circulation (the warming Gulf Stream) can bring sudden and dramatic climate change in the Northern Hemisphere (i.e., paradoxical cooling).
1986	CSIRO Division of Atmospheric Research briefs Australian federal and state Environment Ministers' Council (ANZECC) on the risks posed by the greenhouse effect.
	CSIRO, with support from Australian governments, initiates two conferences—'Greenhouse '87' and 'Greenhouse '88'—that are credited with spurring Australian public understanding of greenhouse to world-leading proportions.
1987	Montreal Protocol of the Vienna Convention achieves international agreement to curb ozone emissions and is cited as an example that international agreement on atmospheric pollution is possible.
	First CSIRO national conference on greenhouse/climate change in Australia.
1988	News coverage of greenhouse effect escalates; framed as risks in response to record heat and drought in the United States and elsewhere. Testimony to US Congress by leading NASA atmospheric scientist James Hansen that he was 99 per cent certain climate change had begun, based on the series of warm years in the 1980s. In Australia, media coverage also in response to second CSIRO and Commission for the Future conference and political/public discourse on the topic.

183

	Intergovernmental Panel on Climate Change (IPCC) established by the World Meteorological Organization (WMO) and the United Nations Environment Programme (UNEP) to advise national governments on best available scientific evidence on climate change causes, consequences, and response strategies, based on peer-reviewed publications; to report to second world climate conference in 1990 (first IPCC report).
1988	Toronto 'Conference on the Changing Atmosphere' attended by scientists, economists, and national leaders; call for action describes human activities as a vast, unplanned experiment upon the planet.
	Level of CO_2 in the atmosphere reaches 350 ppm.
1989	'Declaration of the Hague' by 24 nations including Australia recognises global significance of climate change and calls on all nations to participate in a Framework Convention in 1992.
	Labor federal government under Bob Hawke takes a leading role internationally on climate change.
	April: Federal government sets up a National Climate Change Program with a National Greenhouse Advisory Committee of scientific advisers and a Prime Ministerial Working Group to assess achievable targets, and set priority research areas and objectives.
	The Global Climate Coalition is founded by fossil fuel companies, and other corporations with economic interests in the status quo, to 'fight back' against climate science and proposed action.
1990	First IPCC Assessment Report, made to the second world climate conference in Geneva; Australian scientists play prominent roles on the panel of 170 scientists assessing the published science at this time, backed by 200 scientists conducting peer review of the draft report. First IPCC report notable for its direct and clear language of certainty and risk.

Initiation of ecologically sustainable development (ESD) working groups under Hawke government. A unique attempt to develop sustainable policy in nine sectors of the economy in 'accord' style roundtable discussions by main societal sectors including environmental and community groups, plus government and industry. Greenhouse/climate change tackled by an inter-sectoral group that made significant recommendations, later watered down by federal bureaucracy.

Industry concerns about economic 'cost' of climate change mitigation action enter public discourse; coal industry moves to forefront and 'debate' is initiated.

Federal and state energy portfolio ministers in the Australian Minerals and Energy Council release report, and initiate studies and action to lessen emissions from that sector; significant because it shows early understanding by this portfolio.

October: Federal government releases 'interim planning target' to stabilise CO_2 emissions at 1988 levels by 2000, and reduce them by 20 per cent from there by 2005.

Late 1990 and 1991	Treasurer Paul Keating (elected prime minister in 1991) commissions both ESD greenhouse working group and Industry Commission to investigate cost and benefit of taking action; he receives widely divergent responses; Industry Commission 'frame' focusing on economic cost becomes a pivotal turning point in the national discussion.
1991	Change of federal leadership in Australia, Keating replaces Hawke.
1990s overall	Characterised by increasing influence and then dominance of neo-liberal/free market economic policies, shunning regulation, and shifting from public to privatised energy infrastructure based on coal, gas and hydro-electricity. This period cements investments with 40+ year time span in conventional energy infrastructure and production (e.g., coal-fired electricity plants). Deregulation and competition in energy and other markets switches emphasis from lowering consumer and industrial demand to mitigate emissions, to an emphasis on profit via greater consumption and more supply.
1992	UN Conference on Environment and Development (Rio Earth Summit); Australia still argues for binding emission targets, rejected by the United States under President George H.W. Bush.

1992 UN Framework Convention on Climate Change (FCCC) unveiled at Rio Summit; Australia is a signatory (ratified by federal parliament in December 1992), making it the eighth of 192 parties to have signed by 1994. The convention sets some goals like 2000 as the year for returning emissions to 1990 levels, and obligating signatories to adopt national policies to limit emissions.

National Greenhouse Response Strategy (NGRS) established; reflecting influence of dominant market ideology, NGRS rejects regulation for greenhouse response strategies at federal and state levels. Focus turns to business concerns and priorities, and voluntary industry action, but there is now a reduced focus on alternative energy supply–efficiency measures and renewable technologies at the state level, and international participation at the Commonwealth level.

1994 'Greenhouse '94' organised by CSIRO and New Zealand scientists, organised to review science in lead-up to first conference of the parties to the FCCC. Thereafter, Australian academies of science, engineering and social science report jointly in 1995.

Mid-1990s Scientists gain better understanding of possibilities and mechanisms of rapid climate change; international scientific reports and warnings of risk continue from, inter alia, UK Meteorological Office, the US National Aeronautic and Space Administration (NASA), US National Academy of Sciences, NOAA, and other international institutions.

1995 Second IPCC assessment reports on science, impacts and responses to anthropogenic climate change; confirm and continue the risk analysis set out in 1990 reports; however, language changes to a more cautious/academic modality.

Australian National Greenhouse Response Strategy (NGRS) published but scarcely implemented.

First conference of the parties to the FCCC, held in Berlin, Germany; leads to Berlin Mandate, which calls for agreement, by the end of 1997, on greenhouse gas emission reduction targets. The Kyoto Protocol of 1997 is to establish specifics of targets and methods for each country.

A key frame shift is evident at Berlin; Australia changes its position in international negotiations from ethical-based to an economic-based position, arguing the 'special needs' of its fossil fuel-specialised economy.

1996	Change of federal leadership in Australia to Coalition and John Howard.
	Second conference of the parties to the FCCC held in Geneva, Switzerland. Australia's policy frame continues to shift and Australia establishes itself as a 'climate change laggard' (McDonald 2005: 225).**
	** 'Immediately before the conference the government questioned the science of climate change and opposed the idea of the IPCC's new conclusions on climate change impacts providing the basis for negotiations ... These would "hurt Australia"' (McDonald 2005: 225). Australia was joined by the OPEC states and the Russian Federation. The United States and Europe supported binding emission targets at the time, with the United States under President Bill Clinton who was elected in 1993.
1996–2001	Transition to complete neo-liberal, economic rationalist dominance (hegemony) of public policies and discourses. International stance now about economic 'national interest' and Australia's special case. Cuts or dismantling of research programs focused on energy efficiency, and renewable and alternative sources. Strong ties to neo-liberal think tanks. Attacks on, and marginalisation of, environmentalists. Reports that climate science communication is discouraged from government-funded institutions during this period and later.
1997	Australian Greenhouse Office established. National Greenhouse Advisory Panel (established under Hawke) of scientists, industry, unions, consumers, and government representatives effectively disbanded (not asked to meet after this year). National Greenhouse Response Strategy reviewed; outcome critiqued as weak and ineffective due to lack of leadership, inability to separate public interest from narrow commercial interests, and lack of informed public discourse.
	Australian media reports exhibit strong shift in emphasis from science story to political economic story in the lead-up to the Kyoto Protocol, and document considerable industry resistance to action.
	November: Kyoto Protocol to the FCCC agrees nations to reduce emissions by 2012; signed by 163 countries including Australia (which eventually declined to ratify until a change of government at the end of 2007).
1999	New National Greenhouse Strategy developed with emphasis on voluntary action.

2001	March: Newly inaugurated US President George W. Bush renounces Kyoto Protocol on national emission reduction targets, soon to be joined by Australia; a new stage of political scepticism and denial ramps up in both countries.
	IPCC Third Assessment report; echoes risks outlined in first two assessments in greater regional detail, using language of scientific probability and uncertainty.
	By December 2006 a report by the Australian Greenhouse Office regarding domestic emissions 'predicted greenhouse emissions generated by rising demand for coal-fired electricity would increase by 62 per cent over the next four years, and by 127 per cent by 2020' (Beeby 2006).
2009	Level of CO_2 in atmosphere has risen to 390 ppm. Combined with methane and nitrous oxides (CO_2-e) the level is 450 ppm.
2014	Level of CO_2 in atmosphere measuring above 400 ppm, and scientists voice concern that, without immediate and significant measures to lower global emissions, warming will not be stopped at 2 °C, which is still considered manageable for human societies. Sea level rise of 1–3 metres guaranteed with scientific reports that West Antarctic Ice Sheet has begun irreversible melt. News reports that coastal cities like Miami, United States, already experiencing sea water incursions. In Miami's state of Florida, leading politicians continue to deny the reality of climate change and its effects.
	Australia becomes first country to legislate to undo a national price on carbon pollution and a link to an emissions trading scheme. The carbon price was credited with lower emission measurements after two years in place before it was axed. Australian Government backpedals on successful renewable energy sector.

Sources: Bouma, Pearman & Manning (1996); Diesendorf (2007); Flannery (2005); Hamilton (2001); IPCC (1990, 1995, 2001); Weart (2003); Australian Government documents; industry documents; media reports.

List of acronyms

Australian and New Zealand Environment and Conservation Council (ANZECC)

Australian Broadcasting Corporation (ABC)

Australian Bureau of Agricultural and Resource Economics (ABARE)

Australian Competition and Consumer Commission (ACCC)

Australian Environment Council (AEC)

Australian Industry Greenhouse Network (AIGN)

Australian National Audit Office (ANAO)

Australian Religious Response to Climate Change (ARRCC)

Bureau of Meteorology (BOM)

Business Council of Australia (BCA)

carbon capture and storage (CCS)

Centre for Independent Studies (CIS)

Committee for Economic Development of Australia (CEDA)

Commonwealth Scientific and Industrial Research Organisation (CSIRO)

ecologically sustainable development (ESD)

emission trading scheme (ETS)

Global Warming Policy Foundation (GWPF)

Government Accountability Project (GAP)

Institute of Public Affairs (IPA)

International Geosphere Biosphere Project (IGBP)

National Aeronautics and Space Administration (NASA)

National Greenhouse Response Strategy (NGRS)

National Greenhouse Steering Committee (NGSC)

National Institute of Water and Atmospheric Research (NIWA)

National Oceanic and Atmospheric Administration (NOAA)

National Strategy for Ecologically Sustainable Development (NSESD)

President's Science Advisory Committee (PSAC)

Special Broadcasting Service (SBS)

State Electricity Commission of Victoria (SECV)

Sustainable Energy Development Authority (SEDA)

United Nations Framework Convention on Climate Change (UNFCCC)

United Nations Environment Programme (UNEP)

United Nations Intergovernmental Panel on Climate Change (IPCC)

Union of Concerned Scientists (UCS)

World Economic Degradation General Equilibrium (WEDGE)

Bibliography

ABC (2009, 9 November), *Four Corners*. Retrieved from www.abc.net. au/4corners/content/2009/s2737676.htm.

Ajani, J.A. (2007). *The forest wars*. Carlton, Victoria: Melbourne University Press.

ANAO (1993). *Implementation of an interim greenhouse response—Department of Primary Industries and Energy, Energy Management Programs*. Australian National Audit Office.

Antilla, L. (2005). Climate of scepticism: US newspaper coverage of the science of climate change. *Global Environmental Change* 15, pp. 338–52.

ANZEC (1989). Agenda item 6 (i) Draft national greenhouse strategy. Canberra: Australian and New Zealand Environment Council.

ANZECC (1990). *Towards a national greenhouse strategy for Australia*. Canberra: Australian and New Zealand Environment Conservation Council.

—— (1991). *Report of programs implemented and policies adopted since 1988 that contribute to reducing greenhouse gas emissions in Australia*. Canberra.

Arrangement on nuclear power. *Crikey* (editorial). Retrieved 23 October 2007, from www.crikey.com.

Atmosphere of pressure: political interference in federal climate science (2007, 30 January). *Union of Concerned Scientists*. Retrieved 10 March 2011, from ucsusa.org/scientific_integrity/abuses_of_science/atmosphere-of-pressure. html.

Ashbolt, A. (1987). The ABC in civil society. In E.L. Wheelwright & K.D. Buckley (eds.), *Communications and the media in Australia* (p.115). Sydney, NSW: Allen and Unwin.

Australia Institute (1997). *A policy without a future: Australia's international position on climate change*. No. 8 Background Paper.

Australian science media centre (n.d.). Retrieved 15 January 2011, from www. assmc.org.

Backgrounder on US energy production (1991, February). United States Information Service.

Bagdikian, B. (2004). *The new media monopoly*. Boston, M.A.: Beacon Press.

Barker, M. (2005). *Manufacturing policies: the media's role in the policy making process*. Paper presented at the 'Journalism education' conference. Griffith University.

BBC News (2006, 1 December). 'Maverick' risk to science debate. Retrieved 1 December 2006, from bbc.co.uk/1hi/sci/tech/615937.

Beale, B. (1987, 27 November). A hot spell, forever. *Sydney Morning Herald*, pp. 14–20.

—— (1996, 20 May). Greenhouse gas risk ignored in mine approvals. *Sydney Morning Herald*, p. 9.

Becher, T. (1994). The significance of disciplinary differences. *Studies in Higher Education* 19(2): 151–61.

Beder, S. (1989). *From pipe dreams to tunnel vision: engineering decision-making and Sydney's sewerage system*. PhD thesis, University of NSW.

—— (1999, June). The intellectual sorcery of think tanks. *Arena Magazine* 41: 30–32.

—— (2000). *Global spin: the corporate assault on environmentalism* (1997). Carlton North, Victoria: Scribe.

Beeby, R. (2006, 30 December). Industry snubs climate strategy. *Canberra Times*, p. 1.

—— (2006). Lost in space: where are the conservation groups when we need them? *Canberra Times*, Forum: B2.

—— (2008). Poor showing for science. *Canberra Times*. Retrieved 15 May 2008, from canberra.yourguide.com.au/news/opinion/editorial/general/poor-showing.

Beecher, E. (2005). The decline of the quality press. In R. Manne (ed.), *Do not disturb: is the media failing Australia?* (pp. 169–88). Melbourne: Black Inc.

Benesh, P. (1988, 2 July). Ottawa leads in clearing the air. *Sydney Morning Herald*, p. 18.

Bernays, E. (1928). *Propaganda*. New York: Horace Liverwright.

Bernthal, F.M. (ed.) (1990). *Climate change: the IPCC response strategies*. Washington D.C., Covelo CA: Island Press.

Black, R. (2007). Humans blamed for climate change. Retrieved 6 February 2007, from newsvote.bbc.co.uk/1/hi/sci/tech/632135.

Bogost, I. (2005). *Frame and metaphor in political games*. Paper presented to the DiGRA 2005 Conference: Changing views—worlds in play.

Bolin et al. (1986, 15 October). *The Villach statement: The greenhouse effect, climatic change and ecosystems*. World Meteorological Organisation and United Nations Environment Program.

Bouma, W. & Holper, P. (1990, March). *Climate change—a bibliography of scientific publications from CSIRO Division of Atmospheric Research 1971–1989*. CSIRO.

Bouma, W.J., Pearman, G.I. & Manning M.R. (eds.) (1996). *Greenhouse: coping with climate change*. Collingwood: CSIRO Publishing.

Boyden, S. (1987). *Western civilization in biological perspective*. New York: Oxford University Press.

Boykoff, M.T. & Boykoff, J.M. (2004). Balance as bias: global warming and the US prestige press. *Global Environmental Change-Human and Policy Dimensions* 14(2): 125–36.

—— (2007). Climate change and journalistic norms: a case-study of US mass-media coverage. *Geoforum* 38(6): 1190–204.

Boyle, S. & Ardill, J. (1989). *The greenhouse effect: A practical guide to changing climate*. Hodder and Stoughton.

Brennan, G. & Pincus, J. (2002). Australia's economic institutions. In G. Brennan & G. Castles (eds.), *Australia reshaped: 200 years of institutional transformation* (pp. 53–85). Cambridge University Press.

Breusch, J. (2001, 15 January) Climate change to drive up disaster payouts—insurers. *Australian Financial Review*, p. 16.

Bribes for experts to dispute UN study (2007, 3 February). *Sydney Morning Herald*, p. 1.

Broomhill, R. (2001). Neoliberal globalism and the local state: a regulation approach. *Journal of Australian Political Economy* 48: 117–40.

Broussard, D., Shanahan, J. & McComas, K. (2004). Are issue-cycles culturally constructed? A comparison of French and American coverage of global climate change. *Mass Communication and Society* 7(3): 359–77.

Brown, B. (2008). Garnaut's weak targets recipe for catastrophic climate change, Greens. Retrieved 18 March 2008, from bob-brown.greensmps.org.au/content/media-release/garnauts-weak-targets-recipe-catastrophic-climate-change---greens.

Brown, G.E. Jr., (1996, 23 October). *Environmental science under siege: fringe science and the 104th Congress*. Washington, DC: Democratic Caucus of the Committee on Science, U.S. House of Representatives.

Bubela, T. & Nisbet, M. (2009). Science communication reconsidered. *Nature Biotechnology* 27(6): 514–18.

Buchanan, J. & Tullock, G. (1962). *The calculus of consent*. Ann Arbor: University of Michigan Press.

Bulkeley, H. (2000a). Common knowledge? Public understanding of climate change in Newcastle, Australia. *Public Understanding of Science* 9: 313–33.

—— (2000b). The formation of Australian climate change policy: 1985–1995. In A. Gillespie & W.C.G. Burns (eds.), *Climate change in the South Pacific: impacts and responses in Australia, New Zealand, and small island states* (pp. 33–50). Dortrecht: Kluwer Academic Publishers.

—— (2001). No regrets? Economy and environment in Australia's domestic climate change policy process. *Global Environmental Change* 11(2001), 155–69.

Burton, B. (2014, 8 May). Coal flexes $100 million muscle. *Canberra Times*, Times 2: 4.

Burton, T. (1991, 15 June). Greenhouse bureaucracy is a growing concern. *Sydney Morning Herald*, 32.

Cahill, D. (2004, 29 September). The radical neo-liberal movement and its impact upon Australian politics. Paper presented at the 'Australasian Political Studies' conference. University of Adelaide.

Cahill, D. & Beder, S. (2005). Neo-liberal think tanks and neo-liberal restructuring: learning the lessons from Project Victoria and the privatisation of Victoria's electricity industry. *Social* Alternatives 24(1): 43–48.

Callick, R. (1995, 5 January). Business in last ditch bid to bar carbon tax *Australian Financial Review*, p. 1.

—— (1996a 30 May). Business lines up to fight controls. *Australian Financial Review*, p. 3.

—— (1996b, 4 June). Macquarie fears a greenhouse handicap. *Australian Financial Review*, p. 4.

——(1996c, 12 July). Australia's greenhouse policy wins US support. *Australian Financial Review*, p. 9.

—— (1996d, 5 June) Coalition backs industry on climate change, *Australian Financial Review,* p.2.

Capra, F. (2003).*The hidden connections*. London: Flamingo Harper Collins.

Carey, A. (1987). The ideological management industry. In E.L. Wheelwright (ed.), *Communications and media in Australia* (pp. 158–79). Sydney: Allen and Unwin.

——(1995). *Taking the risk out of democracy: propaganda in the US and Australia*. NSW Press/ Illinois Press.

Carney, S. (1988, 12 September). Most want action over the greenhouse effect. *Sydney Morning Herald*, p. 5.

Carr, K. (2008, 16 January). Liberating the voices of science. *Australian*, p. 23.

Carruthers, D. (1992, April). The earth summit and the risks it poses for Australia's economy. *Mining Review*.

Center for Media and Democracy (2014, 10 July), 'Institute of Public Affairs', *Sourcewatch*. Retrieved from www.sourcewatch.org/index. php?title=Institute_of_Public_Affairs.

Chomsky, N. (1992). *Deterring democracy* (1991). London: Vintage.

—— (1996). *Power and prospects: reflections on human nature and the social order*. Sydney: Allen and Unwin.

Chynoweth, R. (1987). Appropriation Bill (No. 1) 1987–1988. In Committees, House of Representatives, Parliament of Australia. Hansard, p. 961.

Cleary, P. (1990, 5 June). It's the end of the world as we know it. *Sydney Morning Herald*, Magazine: 1.

—— (1991, 26 October). How we can fight our way out of the greenhouse. *Sydney Morning Herald*, p. 34.

Climate change: new Antarctic ice core data. (2000, 20 May). Retrieved 12 April 2011, from www.daviesand.com/Choices/Precautionary_Planning/New_ Data/.

Climate Change Program, Commonwealth n.d.

Climate of distrust. (2005, 7 July). *Nature* (editorial) 436(7047): 1.

Cohen, J. (2006, 13 February). The greenhouse mafia. ABC TV *Four Corners.*

Commission for the Future. (1989). *Personal action guide for the earth.* Canberra: AGPS.

Commonwealth (1990). *Government sets target for reduction in greenhouse gases.* Press release. DASETT and DPIE.

Commonwealth. (1996). *Prospects for Australian industry involvement in greenhouse gas abatement overseas.* Bureau of Industry Economics: Canberra AGPS.

Corbett, J.B., & Durfee, J.L. (2004). Testing public (un)certainty of science: media representations of global warming. *Science Communication* 26(2): 129–51.

Costanza, R., Graumlich, L. & Steffen, W. (eds). (2007). *Sustainability or collapse? Integrated history and future of people on earth.* Dahlem Workshop Report 96. MIT Press.

Crittenden, S. (2006). Common belief on climate change. Transcript. Retrieved 11 November 2006, from old.globalpublicmedia.com/transcripts/813.

Crombie, R. (2003, 21 November). *Inequality and incentive: Don Quixote to mutual obligation.* Paper presented at the 'Social change in the 21st century' conference, Centre for Social Change Research University of Queensland.

Crutzen, P. (2008, 10 September) *The Anthropocene..* Retrieved 10 March 2011, from www.eoearth.org/article/Anthropocene.

Cubby, B. (2008). Climate program swindled viewers. *Sydney Morning Herald.* Retrieved 31 July 2008, from www.smh.com.au/news/global-warming/ climate-program-swindled-viewers/20.

Curtis, A. (2002). *Century of self.* BBC documentary.

Dack, M. (1992, 23 August). *Bureaucratic arrogance weakens national greenhouse strategy.* Media statement. The Institution of Engineers, Australia.

Daley, J. (1992, July*). Australian industry's opposition to a carbon tax.* Paper presented at the CAER 'Energy and the environment' conference.

DaSilva, W. (2008, April/May). Sceptics, denialists, contrarians and delusionists. *Cosmos* (editorial).

Davidson, K. (1992). Defrocking the priests. In D. Horne (ed.), *The trouble with economic rationalism*. Newham, Victoria: Scribe.

Davies, G. (2004). *Economania*. Sydney: ABC Books.

Dayton, L. (1995a, 28 October). It's showdown time for planet earth. *Sydney Morning Herald*, p. 29.

Dayton, L. (1995b, 2 September). Malaria spread linked to climate change. *Sydney Morning Herald*, p. 10.

Delwiche, A. (1995). *Propaganda: why think about propaganda*. Retrieved 10 May 2009, from www.propagandacritic.com/articles/index.html.

Dempster, Q. (2005). The slow destruction of the ABC. In R. Manne (ed.), *Do not disturb: is the media failing Australia?* (pp. 101–20). Melbourne: Black Inc.

Denemark, D. (2005). Mass media and media power in Australia. In S. Wilson (ed.), *Australian social attitudes: the first report* (pp. 221–39). Sydney: UNSW Press.

Department of Arts, Sports, the Environment, Tourism and Territories (1987, 19 October). Climate change due to the greenhouse effect. Memo re. Australian Environment Council meeting.

—— (n.d.). *Climate change program*.

Department of Primary Industries and Energy (1992, February). *Climate Change Newsletter* 4(1).

—— (1996, February). *Climate Change Newsletter* 8(1).

—— (1997, July). *Climate Change Newsletter* 9(1 & 2).

Developing an ACT strategy to respond to the greenhouse effect (1990). Canberra, ACT.

Diamond, J. (2005). *Collapse: how societies choose to fail or survive*. Melbourne: Penguin Group.

Diesendorf, M. (2000). A critique of the Australian Government's greenhouse policies. In A. Gillespie & W.C.G. Burns (eds.), *Climate change in the South Pacific: impacts and responses in Australia, New Zealand and small island states* (pp. 79–93). Dortrecht: Kluwer Academic Publishers.

—— (2007). *Greenhouse solutions with sustainable energy*. Sydney: UNSW Press.

Digsley, J., Deacon, D. & Smart, D. (1979). *Elites in Australia*. Routledge and K. Paul.

Dirty tricks: if science doesn't suit your political viewpoint, suppress it (2007, 3 February). *New Scientist* (editorial) 193(2589): 5.

Dirty tricks: censors exposed (2007, 3 July). *New* Scientist 193(2589): 7.

Dixon, P., Marks, R. & McLennan, P. (1989, September). *The feasibility and implications for Australia of the adoption of the Toronto proposal for carbon dioxide emissions*. CRA Limited.

Doherty, P. (2009, May). Era of complex science: biology and climate. *Australian R&D Review* 9.

Doran, P.T. & Zimmerman, M.K. (2009). Examining the scientific consensus on climate change. *Climate Change* 90(3): 22–23.

Doubting Australia: the roots of Australia's climate denial. Notes prepared by Australian climate organisations seeking action on climate change (2010, May). Retrieved from cana.net.au/sites/default/files/DoubtingAustralia.pdf.

Dryzek, J. (1997). *The politics of the earth: environmental discourses*. Oxford University Press.

Dryzek, J. & Schlosberg, D. (eds.). (1998). *Debating the Earth*. Oxford University Press.

—— (2005). *Debating the Earth* (2nd edn). Oxford University Press.

Dunn, R. (1989, 2 November). Cabinet moves on control of gases. *Australian Financial Review*, p. 8.

Durkin, M. (1997, March). *Against nature*. Channel 4, United Kingdom.

Dwyer, M. (1995, 21 March). Australia takes strong line against greenhouse rules. *Australian Financial Review*, p. 7.

Dwyer M. & Wilson, N. (1995, 6 January). Study argues against $320m carbon tax. *Australian Financial Review*, p. 5.

Dyer, G. & Keane, B. (2013, 21 June). The bespoke, luxury world of *AFR* editor. *Crikey*.

Economics and the environment—the Australian debate (1990). Tasman Economic Research.

Ehrlich, P. & Ehrlich, A. (1998). 'Wise use' and environmental anti-science. In J. Dryzek & D. Schlosberg (eds.), *Debating the Earth* (pp. 70–80). Oxford University Press.

Faine, J. (2005). Talk radio and democracy. In R. Manne (ed.), *Do not disturb: is the media failing Australia?* (pp. 169–188). Melbourne: Black Inc.

Fine, A. (1996). Science made up: constructivist sociology of scientific knowledge. In P. Galison & D.J. Stump (eds.), *The disunity of science*. Stanford University Press.

Flannery, T. (2005). *The Weathermakers*. Melbourne: Text Publishing.

Flint, D. (2003). *The twilight of the elites*. Melbourne: Freedom Publishing Company Pty Ltd.

Fraser, A. (2007). Howard fudges carbon target. *Canberra Times*, p. 1.

Frew, W. (2006, December). Howard's agenda. *The Walkley Magazine* 42: 19–20.

Front groups: the global warming sceptics. *Spinwatch*. Retrieved 20 February 2006, from www.spinwatch.org/modules.php?name=News&file=article&sid=287.

Fyfe, M. (2004). The global warming sceptics. *The Age*. Retrieved 7 December 2007, from www.theage.com.au/articles/2004/11/26/1101219743320.html.

Garnaut, R. (2008). Fateful decisions. *The Garnaut climate change review: final report. Decisions*. Retrieved from www.garnautreview.org.au/pdf/Garnaut_Chapter24.pdf.

Gelbspan, R. (2004). *Boiling point*. New York: Basic Books.

Gilchrist, G. (1995a, 18 October). Australian ploy fails to slow greenhouse action. *Sydney Morning Herald*, p. 1.

—— (1995b, 9 November). Greenhouse effect will cause havoc in NSW, study claims. *Sydney Morning Herald*, p. 5.

—— (1996a, 16 July). Kernot breaks ranks on climate. *Sydney Morning Herald*, p. 1.

—— (1996b, 8 July). Climate changes: why we are seen as rebels. *Sydney Morning Herald*, p. 2.

Gilchrist, G., & Mccathie, A. (1996, 18 July). US climate backflip leaves us in the cold. *Sydney Morning Herald*, p. 1.

Gillis, J. & Chang, K. (2014, 12 May). Scientists warn of rising oceans from polar melt. *New York Times*. Retrieved from www.nytimes.com/2014/05/13/science/earth/collapse-of-parts-of-west-antarctica-ice-sheet-has-begun-scientists-say.html?_r=0.

Gittens, R. (1991, 18 September). Global warming: why we can't act locally. *Sydney Morning Herald*, p. 17.

Glikson, A. (2008a). *RAPID ROUNDUP: Garnaut climate change review: final report — experts react*. Retrieved March 2010, from www.aussmc.org/2008/09/rapid-roundup-garnaut-climate-change-review-final-report-experts-react/#Andrew_Glikson.

—— (2008b). *RAPID ROUNDUP: Is the Earth cooling?—experts respond*. Retrieved 10 March 2010, from www.aussmc.org.au/Is_the_Earth_Cooling.php#Andrew_Glikson.

—— (2009). The lungs of the Earth. Retrieved 10 March 2010, from www.countercurrents.org/glikson021109.htm.

Glover, D. (2005). Is the media pro Labor? In R. Manne (ed.), *Do not disturb: is the media failing Australia?* (pp. 191–215). Melbourne: Black Inc.

Goldberg, M. (2006). Saving secular society *In These Times*. Retrieved 21 May 2006 from www.inthesetimes.com/site/main/article/2649/.

Goldberg, S. (2014, 28 May). Americans care deeply about 'global warming'—but not 'climate change'. Retrieved from www.theguardian.com/environment.

Goldie, J. (2014, July). Australian Climate of 'get behind fossil fuels'. *District Bulletin*, p. 1. Retrieved from www.districtbulletin.com.au.

Greene, D. (1990a). *A greenhouse energy strategy: sustainable energy development for Australia*. Canberra: Department of Arts, Sports, the Environment, Tourism and Territories, AGPS.

—— (1990b). *Reducing greenhouse gases: options for Australia*. ANZEC report 26. Canberra: AGPS.

Greenpeace. (2002). Who's to blame? The role of the US, Canada and Australia in undermining the Rio agreements. Retrieved from www.greenpeace.org/politics/earthsummit/html.

Greenhouse: not just an environmental issue (1996). Australian Coal Association.

Griffiths, T. (1996). *Hunters and collectors: the antiquarian imagination in Australia*. Cambridge University Press.

Griscom-Little, A. (2007a). *And now, a word from our detractor.* Retrieved 1 February 2007, from www.grist.org/news/maindish/2007/01/31/luntz/index.html?source=umbra.

—— (2007b). Thinking outside the Fox. Retrieved 10 May 2007, from www.grist.org/news/maindish/2007/05/09/murdoch/index.html?source=biz.

Gumbel, P. (2009, 2 February). Rethinking Marx: as we work out how to save capitalism, it's worth studying the system's greatest critic. *Time* 173: 39–43.

Hajer, M. (1995). *The politics of environmental discourse, ecological modernisation and the policy process.* Oxford University Press.

Hall, B. (2013, 10 November). Howard is right but it's just a feeling. *Sunday Canberra Times*, p. 16.

Hamilton, C. (2001). *Running from the storm.* Sydney: University of NSW Press.

—— (2006, 20 February). *The dirty politics of climate change.* Paper presented at the 'Climate change and business' conference. Hilton Hotel, Adelaide.

—— (2007). Ratified recalcitrants: our gloss comes off in Bali. Retrieved 10 December 2007, from www.crikey.com.

—— (2010, 29 September). Bullying, lies and the rise of right-wing climate denial. *The drum.* Retrieved from www.abc.net.au/news/2010-02-22/32912.

Hamilton, C. & Maddison S. (2007). *Silencing dissent: how the Australian Government is controlling public opinion and stifling debate.* Sydney: Allen & Unwin.

Hamilton, C. & Throsby, D. (eds.) (1997). *The ESD process: evaluating a policy experiment.* Canberra: Academy of Social Sciences in Australia.

Hannam, P. (2014, 6 May). Wind energy at record levels. *Canberra Times*, p. 5.

Harcher, P. (2007, 1 June). Stealthy moves of the predator. *The Sydney Morning Herald*, p. 11.

Hardin, G. (1968). The tragedy of the commons. *Science* 162: 1243–48.

Harris, S. (1997). The ESD process: background, implementation and aftermath. In C. Hamilton & D. Throsby (eds), *The ESD process: evaluating a policy experiment* (pp. 1–20). Canberra: The Academy of the Social Sciences in Australia.

Hawke, R.J. (1989). *Our country, our future (Statement on the environment).* Canberra: Commonwealth of Australia, AGPS.

——— (1990, 19 March). Speech opening CSIRO Atmospheric Research Building, Aspendale Victoria, press release.

Hayes, C. (2006, November). What we learn when we learn economics. *In These Times* 30: 26–31.

Henderson-Sellers, A. (1990). Australian public perceptions of the greenhouse issue. *Climate Change* 1(69): 96.

——— (1993). An antipodean climate of uncertainty. *Climate Change* 25(3–4): 203–44.

Henderson-Sellers, A. & Blong, R. (1989). *The greenhouse effect, living in a warmer Australia*. Kensington: NSW University Press.

Hengeveld, H. (1997). 1994–95 in review: an assessment of new developments relevant to the science of climate change. *Climate Change Newsletter (DPIE and BRS)* 9(1&2): 1–23.

Holmes, J. (2011, 4 April). It's elementary, my dear Bolt. *ABC Media watch*.

Hordern, N. (2000, 29 March). Greenhouse gas and the high price of hot air. *Australian Financial Review*, p. 18.

Horstman, M. (2005, 23 February). The day after Kyoto. *Catalyst*, program transcript. ABC TV. Retrieved 10 November 2009, from www.abc.net.au/catalyst/stories/s1310331.htm.

Houghton, J.T., Jenkins, G.J. & Ephraums, J. J. (eds.) (1990). *Climate change, the IPCC scientific assessment*. Cambridge, New York, Melbourne: Cambridge University Press.

Houghton, J.T., Meiro Filho, L.G., Callander, B.A., Harris, N., Kattenberg, A. & Maskell, K. (eds). (1995). *Climate change 1995: the science of climate change. Contribution of working group 1 to the second assessment report of the intergovernmental panel on climate change*. Cambridge, New York, Melbourne: Cambridge University Press.

Howard, J. (1997). *Safeguarding the future, Australia's response to climate change*. Statement. Canberra: Commonwealth of Australia.

Howard firm on opposition to Kyoto (2005, October 28). *Canberra Times*, p. 7.

Huck, P. & Macken J. (2001, 2 February). Fossil fools. *Australian Financial Review*, magazine: 34.

Hull, R.B. (2006). God's will and the climate. *New Scientist* 190(2545): 23.

Industry Commission. (1991). *Costs and benefits of reducing greenhouse gas emissions — issues paper*. Canberra: AGPS.

IPCC. (1990, 1995, 2001). Assessment reports. Retrieved from www.ipcc.ch/publications_and_data/publications_and_data_reports.html.

—— *Intergovernmental panel on climate change*. (n.d.) Retrieved 17 January 2011, from www.ipcc.ch.

—— (1990). *Overview and conclusions. Climate change: a key global issue (draft)*.

—— (2007). *Climate change 2007: synthesis report*. Retrieved 17 January 2011, from www.ipcc.ch/publications_and_data/ar4/syr/en/mains5-2.html.

Jacques, P.J., Dunlap, R.E. & Freeman, M. (2008). The organisation of denial: conservative think tanks and environmental scepticism. *Environmental Politics* 17(3): 349–85.

Johnson, C. (2002). *Australian political science and the study of discourse*. Paper presented at the 'Australasian Political Studies jubilee' conference. The Australian National University, 1–4 October.

Johnson, L. (1987). The early years of radio: defining the political. In E.L. Wheelwright & K.D. Buckley (eds.), *Communications and the media in Australia* (pp. 58–77). Sydney: Allen and Unwin.

Jones, B. (1992, 23 March). *Climate change, resource use and population growth: the challenge for sustainable development*: World Meteorological Day 1992 address. Commonwealth Bureau of Meteorology.

—— (2006). *A thinking reed*. Allen and Unwin.

Jones, E. (2002). The ascendancy of an idealist economics in Australia. *Journal of Australian Political Economy* 50: 44–71.

Keeling C. &. Whorf T., (2004, October*)*. *Atmospheric CO_2 from continuous air samples at Mauna Loa observatory, Hawaii, U.S.A*. Carbon Dioxide Information Analysis Center, Oak Ridge National Laboratory. Retrieved 10 March 2011, from cdiac.ornl.gov/trends/co2/sio-keel-flask/sio-keel-flaskmlo_c.html.

Kelly, P. (2009). PM's plan to rescue capitalism. *Australian*, pp. 1, 4.

Kerr, C. (2006, 7 June). A furphy that glows in the dark. Retrieved 8 June 2006, from www.crikey.com.

Keynes, J. (1936). *The general theory of employment, interest and money*. Palgrave Macmillan.

Kinsley, M. (2007, 19 February). Gaffes to the rescue. *Time Magazine*, p. 64.

Klein, N. (2007). *The shock doctrine*. Camberwell, Victoria: Penguin.

Kleinman, R. (2007). A smarter way to cut bills—the weekly greenhouse gas indicator. *Age*, p. 8.

Kolbert, E. (2014). *The sixth extinction—an unnatural history*. London: Bloomsbury.

Kolm J.E. & Walker I.J. (1989). *The potential for reductions of greenhouse gas emissions in Australia*. Prime Minister's Science Council.

Kretschmer, G. (1992, February). ESD greenhouse report. *Climate Change Newsletter* 4(1): 1.

Labor in power (2010). ABC TV documentary.

Laing, R.D. (1971). *The politics of experience* (1967), New York: Ballantyne Books.

Lakoff, G. (2004). *Don't think of an elephant*. Carlton Nth: Scribe Publications.

Lakoff, G. & Johnson, M. (1999). *Philosophy in the flesh: the embodied mind and its challenge to western thought*. New York: Basic Books.

Lavoisier Group. *About the Lavoisier Group*. Retrieved 5 March 2006, from www.lavoisier.com.au/pages/lav-aboutus.html.

Lee, H. & Haites, F. (eds). (1996). *Climate change 1995: economic and social dimensions of climate change, contributions of working group III to the second assessment report of the intergovernmental panel on climate change*. New York: Cambridge University Press.

Leggett, J. (2005). *Half gone: oil, gas, hot air and the global energy crisis*. London: Portobello Books.

Leiserowitz, A., Feinberg, G., Rosenthal, S., Smith, N., Anderson A., Roser-Renouf, C. & Maibach, E. (2014). *What's in a name? Global warming vs. climate change*. Yale University and George Mason University. New Haven, CT: Yale Project on Climate Change Communication.

Lowe, I. (1989). *Living in the greenhouse*. Newham Vic: Scribe Publications.

—— (2006, August). Government determined to silence independent opinion. *Australasian Science* 27(8): 41.

—— (2007). The research community. In C. Hamilton & S. Maddison (eds.). *Silencing dissent* (pp. 60–78). Sydney: Allen and Unwin.

Lowe, T., Brown K., Dessai, S. & de Franca, M. (2006). Does tomorrow ever come? Disaster narratives and public perceptions of climate change. *Public Understanding of Science* 15(4): 435–57.

Luntz, F. (2003). The environment: a cleaner, safer, healthier America. *Straight talk* (pp. 131–46). The Luntz research companies.

Macken, J. (2006, December). Weather or not. *The Walkley Magazine* 42: 15.

Maddox, M. (2005). *God under Howard*. Sydney: Allen and Unwin.

Malone, *P., (2006)* The unabashed rationalist, Peter Boxall. In *Australian department heads under Howard: career paths and practice*. ANU E Press. Retrieved 10 May 2009, from epress.anu.edu.au/anzsog/dept_heads/Mobil.../ch14.html.

Mann, M.E. (2004, 4 December). Myth vs. fact regarding the hockey stick. Retrieved 11 January 2010, from www.realclimate.org/index.php/archives/2004/12/myths-vs-fact-regarding-the-hockey-stick/.

Mann, M.E., & Bradley, R.S. (1999). Northern Hemisphere temperatures during the past millennium: inferences, uncertainties and limitations. *Geophysical Research Letter* 26(6), 759–62.

Mann, M.E., Bradley, R.S. & Hughes, M.K. (n.d.). Northern Hemisphere temperatures of the last six centuries. Retrieved 11 January 2011, from www.ncdc.noaa.gov/paleo/globalwarming/mann.html.

—— (1998). Global-scale temperature patterns and climate forcing over the past six centuries. *Nature* 392: 779–87.

Manne, R. (ed.). (2005). *Do not disturb: is the media failing Australia?* Melbourne: Black Inc.

—— (2011). Bad news: Murdoch's *Australian* and the shaping of the nation. *Quarterly Essay* 43.

Marsh, I. (2005). Opinion formation: problems and prospects. In P. Saunders & J. Walters (eds.), *Ideas and influence* (pp. 220–38). Sydney: UNSW Press.

Martin, P. (2006). Nuclear the way to go: Howard. *Canberra Times*, p. 3.

Massey, M. (1988, 27 July). Government officials start to feel the climate of change. *Australian Financial Review*, p. 28.

Mayne, S. (2006, 2 October). Totally addicted to coal. Retrieved from www.crikey.com.au/articles/2006/02/10-1132-987.print.html.

McDermott, K. (2008). *Whatever Happened to Frank and Fearless? The Impact of New Public Management on the Australian Public Service.* Canberra: ANU E Press.

McDonald, M. (2005). Fair weather friend? Ethics and Australia's approach to global climate change. *Australian Journal of Politics and History* 51(2): 216–34.

McKanna, G. (1988, 30 August). The world (roundup). *Australian Financial Review*, p. 4.

McKenzie, K. (1989, May). Greenhouse professors. *Canberra Times*, Panorama: 33–36.

McKewon, E. (2009). *Resurrecting the war-by-media on climate science: Ian Plimer's Heaven+Earth,* conference paper. Retrieved 22 July 2010, from www.jeaconference.com.au.

McKnight, D. (2005a). Murdoch and the culture war. In R. Manne (ed.), *Do not disturb: is the media failing Australia?* (pp. 53–74). Melbourne: Black Inc.

——— (2005b). *Beyond right and left—new politics and the culture wars.* Allen and Unwin.

McLuhan, M. (1967). *The mechanical bride: folklore of industrial man* (1951). Beacon Press Books (Unitarian Universalist Association).

Mercer, D. (1991). *A question of balance—natural resources conflict issues in Australia.* Annandale NSW: The Federation Press.

Mercer, P. (2006, 24 December). Australia ponders climate future. Retrieved 3 January 2007, from news.bbc.co.uk/1/hi/sci/tech/6204141.stm.

Miller, C. (2001, 24 July). Kyoto lives—just—but Australia may still dump it. *Sydney Morning Herald*, p. 1.

Miller, N. (2013, 6 November). 'The claims are exaggerated': John Howard rejects predictions of global warming catastrophe. *Sydney Morning Herald*. Retrieved from www.smh.com.au/federal-politics/political-news/the-claims-are-exaggerated-john-howard-rejects-predictions-of-global-warming-catastrophe-20131106-2wzza.html.

Mitchell, C.D. (1992). *Grappling with greenhouse.* Canberra National Greenhouse Advisory Committee, CSIRO Publications.

Moffett, I. (1992). *The greenhouse effect, science and policy in the Northern Territory*, North Australian Research Unit, The Australian National University.

Monbiot, G. (1997). *The revolution has been televised*. Retrieved 4 February 2009, from www.monbio.com/archives/1997/12/18/the-revolution-has-been-televised/.

—— (2005, 30 June). Going nowhere. *Guardian*.

Mooney, C. (2005). *The Republican war on science*. New York: Basic Books

Moran, A. & Chisholm, A. (1991). *Greenhouse gas abatement: its costs and practicalities*: Tasman Institute.

Morgan, R. (1989). Public perception and attitudes towards 'the greenhouse effect' —including their relation to energy consumption. Electricity Commission of NSW.

Mother nature's revenge: The 'greenhouse effect' and ozone depletion threaten major changes in our health and climate (1987, 2 March). *Newsweek*, p. 1.

Mounser, L. (2000, 19 October). Hot news the greenhouse effect is not so bad after all. *Sydney Morning Herald*, p. 12.

Murray, G. & Pacheco, D. (2000). The economic Liberal ideas industry: Australasian pro-market think tanks in the 1990s. *Journal of Social Issue*, May.

Murphy, K. (2006, 30 December). PM puts faith in nuclear power. Retrieved 12 December 2011, from www.theage.com.au/news/national/pm-puts-faith-in-nuclear-power/2006/12/29/1166895479257.html.

National Research Council (2006). *Committee on surface temperature reconstructions for the last 2,000 years report*. Washington, D.C.

Nahan, M. (2003, December). The demise of science. *IPA Review*.

Newby, J. (2005, 24 November). Real oil crisis. ABC TV *Catalyst*.

New South Wales in climate strategy (1989, 18 July). *Australian Financial Review*, p. 3.

Nichols, J. & McChesney, R. (2005). *Tragedy and farce: how the American media sell wars, spin elections, and destroy democracy*. New York: The New Press.

Nisbet, M. & Myers, T. (2007). The polls—trends, twenty years of public opinion about global warming. *Public Opinion Quarterly* 71(3): 444–70.

Nisbet, M. & Mooney, C. (2007, 31 August). Framing science. *Science*, pp. 1168–70.

Ocean current shutdown (2009). Retrieved 14 January 2011, from www.sciencedaily.com/releases/2009/07/090716141142.html.

Ophuls, W. & Boyan, S. (1998). The American political economy II: the non-politics of laissez faire. In J. Dryzek & D. Schlosberg (eds.), *Debating the Earth* (pp. 187–202). Oxford University Press.

Oreskes, N. (1999). *The rejection of continental drift: theory and method in American earth science.* New York: Oxford University Press.

—— (2004a, 3 December). The scientific consensus on climate change. *Science* 306: 1686.

—— (2004b). Science and public policy: What's proof got to do with it? *Environmental Science and Policy* 7: 369–83.

Oreskes, N. & Conway, E.M. (2012). *Merchants of doubt.* New York: Bloomsbury Press.

Palfreman, J. (2006). A tale of two fears: exploring media depictions of nuclear power and global warming. *Review of policy research* 23(1): 23–43.

Paltridge, G. (2004). The politicised science of climate change. *Quadrant Magazine* 48(10): 14–18.

Pearce, F. (1989). *Turning up the heat.* London: Paladin Books (Collins).

—— (2005, 15 October). The week the climate changed. *New Scientist* 188(2521): 52–53.

Pearman, G.I. (ed.) (1980). *Carbon dioxide and climate—Australian research.* Canberra: Australian Academy of Science.

—— (ed.) (1988). *Greenhouse: planning for climate change.* CSIRO/Brill.

Pearse, G. (2005). The business response to climate change: case studies of Australian interest groups. PhD thesis, The Australian National University.

—— (2007). *High and dry.* Camberwell: Penguin Vintage.

—— (2009). Quarry vision: coal, climate change and the end of the resources boom. *Quarterly Essay*, 33.

Pearson, C. (2009, 18 April). Sceptic spells doom for alarmists. *The Australian.* Retrieved from www.theaustralian.com.au/archive/news/sceptic-spells-doom-for-alarmists/story-e.

—— (2006, 10 June) Rising tide of bad science, *The Weekend Australian,* p. 28.

Peating, S. (2007, 3–4 February). World wakes up to climate hell. *Sydney Morning Herald*, p. 1.

Petit, J. & Raynaud, D. (1999). 420,000 years of atmospheric history revealed by the Vostock ice core, Antarctica. Press release. Retrieved 12 May 2011, from www.cnrs.fr/cw/en/pres/compress/mist030699.html.

Pittock, A.B. (1987a). Climate catastrophes. Peace Research Centre.

—— (1987b). Forests beyond 2000—effects of atmospheric change. *Australian Forestry* 50(4): 205–15.

—— (1988). Actual and anticipated changes in Australia's climate. In G. Pearman (ed.), *Greenhouse, planning for climate change* (pp. 25–51). CSIRO/Brill.

—— (1991). Climate scenarios for 2010 and 2050 AD Australia and New Zealand. *Climate change* 18: 259–69.

Plimer, I. (2007). The past is the key to the present: greenhouse and icehouse over time. Retrieved 7 December 2007, from www.ipa.org.au/publications/publishing _detail.asp?pubid=196.

Pockley, P. (2007, April). Global warming: worse than forecast. *Australasian Science* 28: 28–31.

'Providing insight into climate change: 61 scientists request a review of climate science'. *Friends of science*. Retrieved from www.friendsofscience.org/index.php?id=113.

Pusey, M. (1991). *Economic rationalism in Canberra*. Cambridge University Press.

Quiddington, P. (1988, 23 August). Scientists warn of islands' peril. *Sydney Morning Herald*, p. 7.

Quiggin, J. (2005). Economic liberalism: fall, revival and resistance. In P. Saunders & J. Walter (eds.), *Ideas and influence*. Sydney: Univeristy of NSW Press.

—— (2006, December). Say it isn't so …. *The Walkley Magazine*, p. 16.

Quinn, N. (1987, 19 October). *Climate change due to the greenhouse effect*. Memo. Department of Arts, Sports, the Environment, Tourism and Territories.

Rampton, S. & Stauber, J. (2002). *Trust us, we're experts*. New York: Tarcher/Putnam.

Real climate: climate science from climate scientists. (n.d.) Retrieved 15 January 2010, from www.realclimate.org.

Reinecke, I. (1987). Information and the poverty of technology. In E.L. Wheelwright & K.D. Buckley (eds.), *Communications and the media in Australia*. Sydney: Allen and Unwin.

Restoring scientific integrity in policy making. (2004). Union of Concerned Scientists (UCS). Retrieved 14 April 2007, from www.ucsusa.org/scientific_integrity/abuses_of_science/investigations_and_surveys/reports-scientific-integrity.html.

Revkin, A. (2013, 12 October). Digital and social media in the West. *Slideshare*. Retrieved from www.slideshare.net/Revkin/talk-china-yale-final-10-1113.

Rignot, E., Mouginot, J., Morlighem, M., Seroussim, H. & Scheuchl, B. (2014, 27 May). Widespread, rapid grounding line retreat of Pine Island, Thwaites, Smith, and Kohler glaciers, West Antarctica, from 1992 to 2011. *Geophysical Research Letters* 41(10): 3502–09.

Rinehart, G. (2013, 18 April). Rupert Murdoch's outstanding address. *Quadrant online*. Retrieved from quadrant.org.au/opinion/qed/2013/04/rupert-murdoch-s-outstanding-address/.

Roberts, P. (1989, 21 December). Energy reappraisal overdue. *Australian Financial Review*, p. 17.

Rosenbloom, H. (1979). *Politics and the media*. Melbourne: Scribe Publications.

Roskam, J., Berg, C. & Paterson, J. Be like Gough: 75 radical ideas to transform Australia. *IPA Review*, p. 2. ipa.org.au/publications/2080/be-like-gough-75-radical-ideas-to-transform-australia.

Ruddiman, B. (2005, March). How did humans first alter global climate? *Scientific American*. Retrieved 6 January 2010, from www.realclimate.org/index.php/archives/2005/12early-anthropocene-hypothesis.

Rundle, G. (2005). The rise of the right. In R. Manne (ed.). *Do not disturb: is the media failing Australia?* Melbourne: Black Inc.

Russell, C. (2006). *Covering controversial science: improving reporting on science and public policy*. Politics and Public Policy report #2006-4. John F Kennedy School of Government, Harvard University.

Sachs, J. (2009, 19 January). The case for bigger government. *Time*, 173: 20–22.

Saddler, H. (1996). Greenhouse policies and the Australian energy supply industries. In W.J. Bouma, G.I. Pearman & M.R. Manning (eds.). *Greenhouse: coping with climate change*. Collingwood: CSIRO Publishing.

Salinger, J. (2014, 23 January). An insider's story of the global attack on climate science, *theconversation.com*.

Sanderson, W. (2006). Time for the media to get serious about climate change. Retrieved 31 May 2006, from www.crikey.com.au/2006/05/29/time-for-the-media-to-get-serious-about-climate-change/.

Sandilands, B. (2008). Rudd's stunted vision. *Commentary wrap: the pundits react to Rudd's ETS*. Retrieved 21 December 2008, from www.crikey.com.

Sarewitz, D. (2004). How science makes environmental controversy worse. *Environmental Science and Policy* 7(5): 385–403.

Schauble, J. (2001, 23 January). Six degrees hotter: global climate alarm bells ring louder. *Sydney Morning Herald*, p. 1.

Schneider, S. (1976). *The genesis strategy*. New York and London: Plenum Press.

—— (1989). *Are we entering the greenhouse century?* Cambridge: Lutterworth Press.

Schubert, M. (2006, 4 October). Howard names his three towering heros. *The Age*, p. 1.

Science and public policy institute, about. Retrieved 7 December 2007 from scienceandpublicpolicy.org/about_us.html.

Scott, S. (2000, 14 September) A shrinking threat. *Australian Financial Review*, p. 34.

Seccombe, M. (1988, 28 September). Environment beyond our generation. *Sydney Morning Herald*, p. 17.

—— (1990a, 6 September). Polluters put on the back-burner. *Sydney Morning Herald*, p. 1.

—— (1990b, 9 October). Too much hot air and not enough energy. *Sydney Morning Herald*, p. 15.

SECV (1989). *The SEC and the greenhouse effect*. Victoria.

Senate Standing Committee on Industry, Science and Technology (1989, 21 December). *The contribution that Australian industry, science and technology can make to reducing the impact of the greenhouse effect*. Commonwealth Canberra.

Self-interest key to weather debate (2014, 16 January). *Canberra Times*, Times 2: 2.

Shanahan, A. (2004, 24 July). Roo blues. *Canberra Times*.

Shears, R. (1980, February). The greenhouse syndrome. *Playboy*, pp. 96–102.

Sheehan, P. (1988, 24 June). Greenhouse—science's nightmare now reality. *Sydney Morning Herald*, p. 1.

Shellenberger, M. & Nordhaus, T. (2005). The death of environmentalism. Retrieved from www.grist.org/news/maindish/2005/01/13/little-doe.

Shiller, H.I. (1992). *Mass communications and American empire* (1969). Westview Press.

Singer, P. (1993). *How are we to live? Ethics in an age of self interest*. Text Publishing Company: Melbourne.

Smith, B. (2011, 12 January). Ocean temperatures exacerbate la niña. *Canberra Times*, p. 3.

Smith, D. (2000, 13 November). Pointing the finger at the main villain. *Sydney Morning Herald*, p. 13.

Smythe, D.W. (1981) *Dependency road: communications, capitalism, consciousness and Canada*. Ablex Pub Corp.

Steffen, W. (2007). *The anthropocene: from hunter-gatherers to global geophysical force* –Powerpoint Presentation.

Steffen, W., Crutzen, P.J. and McNeill, J.R. (2007). The anthropocene: are humans now overwhelming the great forces of nature? *Ambio* 36: 614–21.

Steketee, M. (2007, June 7). Green row will be decided on economic fear. *Australian*, p. 10.

Stewart, J. (2007, June 4). The real climate challenge. *Canberra Times*, p. 17.

Still in a mess over climate change (2006, 30 September). *New Scientist* (editorial) 191(2571): 5.

Stone, J. (1996, 13 June). Doomsaying die-hards. *Australian Financial Review*, p. 25.

Strangio, P. (2006, September). Lecture to the Australian Senate. *The Public Sector Informant* 15.

Strong, G. (2005, 31 October). It's climate change, as forecast. *The Age*. Retrieved 3 March 2011, from www.theage.com.au/news/opinion/its-climate-change-as-forecast/2005/10/30/1130607148560.html.

Sturgess, G. & Torrens, N. (2009, 21 July). *Liberal rule: the politics that changed Australia*. SBS documentary.

Stutchbury, M. (1990, 7 March). Australia's greenhouse dilemma disregarded by election campaign. *Australian Financial Review*, p. 16.

—— (1995, 17 February). Clash of principles in greenhouse debate. *Australian Financial Review*, p. 25.

Surface temperature reconstructions for the past 2,000 years (2006). Committee on Surface Temperature Reconstructions, f.t.P., 2,000 Years. Washington, D.C.: The National Academies Press.

Sutton, P. (2006, June). One part of Australia that isn't going nuclear. Retrieved 9 June 2006, from www.crikey.com.au/2006/06/07/one-part-of-australia-that-isnt-going-nuclear/.

Taylor, L. (2000, 11 April). Businessmen throw stones at greenhouse. *Australian Financial Review*, p. 4.

Thacker, P. (2006). Climate skeptics in Europe? Mostly missing in action. Retrieved 23 July 2006, from www.sej.org/pub/SEJournal Excerpts.htm.

The changing atmosphere: implications for global security. (1988, 27 June). Conference statement, Environment Canada. Toronto, Canada.

The global climate change lobby (n.d.). Retrieved 12 May 2009, from www.publicintegrity.org/investigations/global_climate_change_lobby/.

Tiffen, R. (1989). *News and power*. Sydney: Allen and Unwin.

Tucker, B. (1994, 28 November). Greenhouse: facts and fancies. *IPA Environmental Backgrounder*, pp. 1–10.

Tuler, S. (1998) *The politics of the Earth: environmental discourses*. Retrieved 10 March 2009, from www.humanecologyreview.org/pastissues/her51/51bookreviews.pdf.

Union of Concerned Scientists (2004). *Scientific integrity in policy making*. Retrieved 10 April 2004, from www.ucsusa.org/scientific_integrity/abuses_of_science/investigations_and_surveys/reports-scientific-integrity.html.

Understanding climate change: a program for action (1975). Washington D.C.: National Academy of Sciences.

Ungar, S. (2000, July). Knowledge, ignorance and the popular culture: climate change versus the ozone hole. *Public Understanding of Science*, 9(3): 207–312.

US and Australia confirm opposition to Kyoto (2006, November 30). *Australian Business Council for Sustainable Energy CO$_2$ News*. Retrieved 2 December 2006, from news.envirocentre.com.au/co2/newsletter.php.

Walker, I.J. (1996). Carbon dioxide from the Australian energy sector: an energy-efficient scenario. In W.J. Bouma, G.I. Pearman & M.R. Manning (eds.), *Greenhouse: coping with climate change*. Collingwood: CSIRO Publishing.

Walsh, Bryan (2013, 25 November). The typhoon's toll. *Time*, p. 22.

Ward, I. (2001). *Politics of the media* (1995). South Yarra: Macmillan.

Warren, M. (2007, 3–4 February). It's almost certain: humans caused planet to heat up. *Weekend Australian*, p. 1.

Weart, S. (2003, August). The discovery of rapid climate change. Retrieved 15 July 2006, from www.physicstoday.org/vol-56/iss-8/p30.html.

—— (2004). *The discovery of global warming* (2003). Cambridge: Harvard University Press.

Weingart, P., Engels, A. & Pansegrau, P. (2000). Risks of communication: discourses on climate change in science, politics and the mass media. *Public Understanding of Science*, 9(3): 261–83.

Wheelwright, E.L. & Buckley, K.D., (eds.) (1987). *Communication and the media in Australia*. Sydney: Allen and Unwin.

White, R.M. (1990). The great climate debate. *Scientific American* 263(1): 18–25.

Wilkenfeld G., Hamilton, C. & Saddler, H. (1995, February). *Australia's greenhouse strategy: can the future be rescued?* Discussion paper no. 3. The Australia Institute.

Wilkinson, M., (2007, March 24). Delayed reaction. Retrieved 9 January 2009, from www.smh.com.au/news/environment/delayed-reaction/2007/03/23/11745978827.

Wilson, E.O. (2005, September). A brave new world. *Cosmos — the science of everything*, pp. 64–69.

Wilson, K.M. (2000). Drought, debate, and uncertainty: measuring reporters' knowledge and ignorance about climate change. *Public Understanding of Science* 9: 1–13.

Wilson, N. (1995, 16 January). Tax incentives preferred. *Australian Financial Review*, p. 8.

Windram, C. (2000, 4 May). Sea change in eco-policy gathers pace. *Australian Financial Review*, p. 23.

World Meteorological Organisation (WMO) (1986) *Report of the International Conference on the assessment of the role of carbon dioxide and of other greenhouse gases in climate variations and associated impacts.* Villach, Austria, 9–15 October 1985, WMO No.661.

Wright, R. (2004). *A short history of progress.* Melbourne: The Text Publishing Company.

Yager, R. E. (1991). The constructivist learning model: towards real reform in science eduction. *The Science Teacher*, 58(6): 52–57.

Zillman, J. (2004). Climate change: a natural hazard? Retrieved 2 February 2009, from www.lavoisier.com.au/articles/greenhouse-science/climate-change/zillman2004-8.php.

—— (2006). The intergovernmental panel on climate change. Retrieved 10 November 2009, from www.atse.org.au/index.php?sectionid=790.